SH
156.9
.R67
1999

Anaesthetic and Sedative Techniques for Aquatic Animals

BELL LIBRARY
TEXAS A&M UNIVERSITY
CORPUS CHRISTI

Anaesthetic and Sedative Techniques for Aquatic Animals

Lindsay G. Ross, BSc, PhD
Barbara Ross, BSc, PhD

Institute of Aquaculture
University of Stirling
Stirling FK9 4LA

Second Edition

Blackwell
Science

© 1999 by
Blackwell Science Ltd
Editorial Offices:
Osney Mead, Oxford OX2 0EL
25 John Street, London WC1N 2BL
23 Ainslie Place, Edinburgh EH3 6AJ
350 Main Street, Malden
MA 02148 5018, USA
54 University Street, Carlton
Victoria 3053, Australia
10, rue Casimir Delavigne
75006 Paris, France

Other Editorial Offices:

Blackwell Wissenschafts-Verlag GmbH
Kurfürstendamm 57
10707 Berlin, Germany

Blackwell Science KK
MG Kodenmacho Building
7–10 Kodenmacho Nihombashi
Chuo-ku, Tokyo 104, Japan

The right of the Author to be identified as the Author of this Work has been asserted in accordance with the Copyright, Designs and Patents Act 1988.

All rights reserved. No part of this publication may be reproduced, stored in a retrieval system, or transmitted, in any form or by any means, electronic, mechanical, photocopying, recording or otherwise, except as permitted by the UK Copyright, Designs and Patents Act 1988, without the prior permission of the publisher.

First published 1984
Second edition published by
Blackwell Science Ltd 1999

Printed and bound in Great Britain by
MPG Books Ltd, Bodmin, Cornwall

The Blackwell Science logo is a trade mark of Blackwell Science Ltd, registered at the United Kingdom Trade Marks Registry

DISTRIBUTORS

Marston Book Services Ltd
PO Box 269
Abingdon
Oxon OX14 4YN
(*Orders:* Tel: 01235 465500
Fax: 01235 465555)

USA
Blackwell Science, Inc.
Commerce Place
350 Main Street
Malden, MA 02148 5018
(*Orders:* Tel: 800 759 6102
781 388 8250
Fax: 781 388 8255)

Canada
Login Brothers Book Company
324 Saulteaux Crescent
Winnipeg, Manitoba R3J 3T2
(*Orders:* Tel: 204 837-2987
Fax: 204 837-3116)

Australia
Blackwell Science Pty Ltd
54 University Street
Carlton, Victoria 3053
(*Orders:* Tel: 03 9347 0300
Fax: 03 9347 5001)

A catalogue record for this title is available from the British Library

ISBN 0-632-05252-X

For further information on Blackwell Science, visit our website:
www.blackwell-science.com

Contents

Preface to the First Edition (1984)

The purpose of this handbook is to draw together the available information on sedation and anaesthesia of fish. Both temperate and tropical freshwater species are considered as well as sedation in sea water. While sedation is a routine and essentially simple procedure, it can also be mismanaged. The overall intention is therefore to produce an illustrated, practical guide for workers both in aquaculture and research.

Acknowledgements

This handbook is one of a series produced by the Institute of Aquaculture of the University of Stirling to provide practical, readily available sources of information on various aspects of temperate aquaculture.

Some information presented here was generated while the authors were involved with the tropical aquaculture programme supported at Stirling by the Overseas Development Administration. Much of the remaining data are based on our own experience and that of many of our colleagues at Stirling over the last 10 years and we are grateful to all concerned for useful discussions and collaboration. Thanks are also due to Lyn North for the cover graphics.

Preface to the Second Edition

In 1983 we were invited to a meeting of the UK Veterinary Anaesthetists Association in Edinburgh to talk to an experienced group of researchers about fish anaesthesia, a minor and unusual topic for most veterinary anaesthetists. Both of us had been involved for some time with fish and invertebrate research requiring a range of anaesthetic approaches and had been teaching fish anaesthesia in various courses in the Institute of Aquaculture for some years. The simplest thing to do in this circumstance was to turn our lecture notes into an illustrated talk, which we did. Based upon this, we wrote the first edition of this handbook in 1983 for publication in the series of handbooks which the Institute of Aquaculture intended to produce at that time. The basic idea was to draw together the available information on sedation and anaesthesia of fish. Both temperate and tropical freshwater species were included, as well as sedation in sea water. We recognised that sedation and anaesthesia can be mismanaged and so the overall intention was to produce an illustrated, practical guide for workers both in aquaculture and research which was accessible and as cheap as possible.

Fifteen years have passed since the handbook first appeared. It sold out long ago and yet is still in demand. The major part of this second edition was prepared in 1994/1995, with selected additions up to 1998. It brings the content up to date, as much of significance has been published in the last decade. In the course of researching the literature, in excess of 1000 new articles or monographs have been unearthed. Having said that, this book is not intended to be an exhaustive review of aquatic animal anaesthesia. Its contents are selective and are firmly aimed at providing an essentially practical reference for the laboratory or the farm. The references provided are also selective, but should be sufficient to provide the reader with a starting point in the extensive literature for following up a topic.

Some of the introductory sections to the new edition have been expanded to give more detailed explanations of mechanisms involved and to provide a better background for those who wish to read a little about them. Sections on anaesthesia of invertebrates and amphibians and reptiles are now included, as these phyla are of importance in aquaculture. New descriptive sections on most drugs used for anaesthesia of aquatic animals have been included and some novel anaesthetic drugs are also dealt with. A chapter on transportation has been included, as calming and sedative techniques have a useful role in this important operation. Glossaries have been included giving chemical details of drugs and an explanation of major technical terms. The index added to this expanded edition will also be useful.

Much of the information reported here is based upon our own experience, general practice and the extensive literature and was up-to-date at the time of writing. However, anaesthesia and sedation is just part of the aquatic biologist's toolbox, and if mismanaged it can ruin everything. There are many disasters, unreported of course, which occur as workers rush headlong into untested methods. We consequently urge extreme caution and judicious pre-experimentation upon all whom intend to carry out animal sedation or anaesthesia, of any kind. In recent years, the legislation covering procedures and materials in this book has been steadily improved and extended. All users of the techniques described here should be aware of the relevant legislation on animal welfare, safe operator practice for drugs and chemicals and their storage, safety legislation concerning electric fishing and similar electrical apparatus, use of drugs with animals intended for the human food chain and release of drugs into the environment.

The authors and the publisher accept no responsibility for losses of animals, losses of experimental or other data, harm occurring to individuals or any other matter which may arise caused by application of the data contained herein or which may arise as a result of non-compliance with any relevant legislation.

In attempting to reach a wide readership, we have tried to provide sufficient background while maintaining a strong practical element in the book. We hope that this is a book not only for the office shelf, but also for the lab, field, or farm and that readers will continue to find it approachable.

Lindsay G. Ross
Barbara Ross

Acknowledgements

In the first edition we invited readers to let us know of their problems and experiences. We are grateful to all those who have indeed done so as this has kept us in touch with a wide range of activities world-wide which would not otherwise be reported in the literature. Thank you for your many letters. We also wish to acknowledge our students, David McCaldon, Remigius Okoye, Elizabeth Robinson and Bernard Ladu, who worked directly in this field with us over the years and our many colleagues, particularly Dr Malcom Beveridge, who supplied us with interesting snippets of data gathered from all over the world

The major part of this revision was written while LGR was on sabbatical leave in Mexico and sincere thanks are due to my old friend Dr Carlos Martinez-Palacios, Director of CIAD, Unidad Mazatlan en Acuicultura y Manejo Ambiental, for providing me with the opportunity to do some thinking in peaceful surroundings and to "break the back" of the work during 1994. I am also grateful to Dr Innocencio Higuera-Ciapara, Director General of CIAD, Hermosillo, Mexico, for his support and generosity during my stay in Mexico. My thanks are also due to my research student, Dr Pepe Aguilar, and to Julia Farrington for sending me the many small items I had forgotten, or could not obtain, in Mexico.

My special thanks go to my wife Barbara and my daughter Bryony Louise, who tolerated my absence during 1994. Since the first edition, Dr Barbara Ross has departed for a new career in teaching. She has, however, contributed to this book in great measure, not only by allowing LGR to spend sabbatical time in Mexico, but also by advising on content and formatting.

Lindsay G. Ross

Chapter 1
Introduction

The handling of aquatic animals both in and out of their natural environment almost always creates great difficulties. Their characteristic struggling during capture and handling usually has strong effects on both physiology and behaviour, and consequently it is often necessary to immobilise fish before attempting to perform even the most simple task (Tytler and Hawkins, 1981). Anaesthesia and sedation are valuable tools in wild fish collection and for fisheries management. In the operation of a culture facility it is not frequently necessary to sedate or anaesthetise stocks. For certain procedures, however, sedation may be essential to minimise stress or physical damage caused by crowding, capture, handling and release and in research work or veterinary practice there may be the additional requirements to render animals unconscious or to alleviate pain.

"Comfort" and animal husbandry

The principal objectives of aquaculture are to grow animals to market size in the minimum of time, with minimum inputs and with minimum cost. This implicitly means that both growth rate and growth efficiency must be maximised and, for this, animals must be unstressed and "comfortable" within their culture environment. In writing about physiological choices in the natural environment, Balchen (1975) suggested that animals would select, or attempt to select, environments in which they were able to "maximise their comfort". This concept can easily be grasped if one considers, for example, the selective movements that a fish would make when presented with a temperature or salinity gradient. A fish would swim into an area of lower or higher temperature which it "preferred", perhaps in the process optimising the temperature for some physiological process such as feeding or digestion, but inevitably choosing an area of best "comfort". This ability to choose applies to a wide range of conditions and situations that may be encountered daily, or seasonally, in the life of an aquatic animal. In the more managed and often confined environments of aquaculture, animals are unable to select their environment in the same way and it is then the task of the aquaculturist to provide an environment in which the cultured animals are as stress-free as possible, and hence "comfortable".

Aquatic animals in extensive culture are relatively unstressed, but in more intensive systems they are, almost certainly, in a state of chronic low-level stress, caused by the combined effects of some kind of confinement, high culture density, behavioural disruption and possibly compromised environmental factors. These environmental factors can be wide ranging and complex and include poor, or partly degraded, water quality, over-simplistic system design, inadequate provision of biologically meaningful systems and practices and, generally, sub-optimal animal husbandry. Such problems are often the result of trade-offs in design and costs, which lead to convenient and economical systems in terms of their operation and management but which nevertheless are very different from conditions in the natural environment. Little is known of the real extent of this problem in aquaculture, or how it is manifested in different systems and in different species, but many workers now feel that chronic low-level stress is a real and probably constant constraint. Clearly, good animal husbandry can minimise its damaging effects, although good husbandry may simultaneously tend to suppress or minimise any tell-tale clinical signs which would normally be apparent. Any additional handling, or similar operation, will cause extra, acute stress which will leave animals in a weakened state, perhaps feeding badly for some time, with poor digestive efficiency and hence growth, greatly predisposed to disease and much less able to deal with environmental fluctuations.

Handling and mechanical damage

There are many instances where some form of restraint or calming is needed to facilitate handling. In struggling aquatic animals there is the additional, and very significant, effect of gross physical damage – a problem which is not usually encountered in handling terrestrial species, at least not in the same way. The skin of fish is normally very thin and is external to the scales, making it very easy to abrade. The delicate eyestalks and appendages of crustaceans are also particularly vulnerable and great care must be taken to avoid the damaging abrasion, or even possible mutilation, which may occur during any handling process. It must be stressed that, even where animals are not being handled individually, but are involved in bulk operations such as batch weighing, there will be substantial risk of abrasion. This is a factor that is often overlooked in practice, especially where procedures are brief, in the belief that brevity enables evasion of consequences.

Pain

In some areas of aquaculture and in research or veterinary work there may be the additional need to address the problem of pain. The ability of fish and invertebrates to feel pain is frequently contended. However, it is clear that vertebrate animals do react to noxious stimuli and it is well known that most invertebrates also have special receptors for such stimuli. Indeed, because specific receptors (nociceptors) and their axons can be readily identified in invertebrates, they have been widely used as models in pain research for many years. Any procedure which involves invasive methods (ranging from fairly simple forms of tagging using needles, to any use of incisions and suturing or to internal surgery) is certain to cause some degree of pain, particularly in the vertebrates but probably also to some degree in the invertebrates, and steps must be taken to alleviate any suffering which could be caused. The humane treatment of experimental or cultured animals must take precedence over other considerations (Green, 1979).

In summary

When aquatic animals are removed from water, individually or as groups, physiological stress is compounded by the risk of serious abrasion and mechanical shock, particularly with a struggling animal. Procedures which can be carried out entirely in water, such as simple tagging, grading using screens or pumped water systems and batch weighing in water, are unlikely to cause serious harm. For example, experienced workers know that careful but firm handling of brood-stock trout can enable procedures such as stripping of eggs and sperm without sedation. However, there are many instances in which some form of calming technique, or sedation, will be required, for example during length-weight studies, sexing, simple injection, withdrawal of body fluids, branding and certain types of tagging. In other cases additional time may be needed to handle animals or carry out surgery, for example during gonadectomy, hypophysectomy, tissue biopsy or prolonged physiological investigations, and then full surgical anaesthesia will be required.

It should be noted that all sedative and anaesthetic procedures themselves induce side effects which may not be considered desirable. For example, Allison (1961) has shown that sperm motility was greatly reduced in brook trout at 19 $mg.l^{-1}$ MS222 and clearly it is inadvisable to expose gametes to anaesthetics. Overall, however, their advantages generally outweigh their disadvantages if the correct technique is used and if sufficient control is maintained.

References

Allison, L.N. (1961). The effect of tricaine methanesulphonate (MS222) on the motility of brook trout sperm. *Prog. Fish. Cult.* 23: 46-48.

Balchen, J.G. (1976). Modelling of the biological state of fishes. SINTEF: The engineering research foundation of the Technical University of Norway, Trondheim. Teknical note 62 for NTNF/NFFR. 25pp.

Green, C.J. (1979). *Animal Anaesthesia*. Laboratory Animal Handbooks. No. 8. Laboratory Animals Ltd, London. 300pp.

Tytler, P. and Hawkins, A.D. (1981). Vivisection, anaesthetics and minor surgery. In: *Aquarium Systems* (Edited by A.D. Hawkins). Academic Press, London.

~~~~

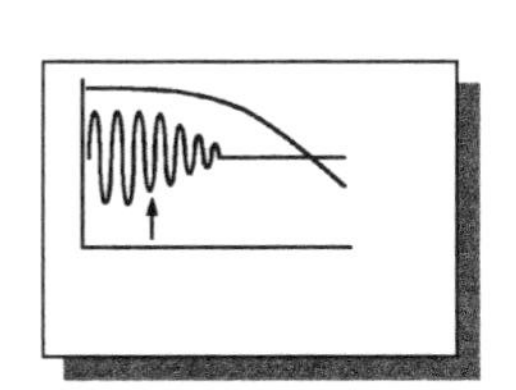
~~~~

Chapter 2
Stress in Aquatic Animals

Introduction

Almost any stimulus presented to an animal causes one or more of a number of behavioural and physiological changes. Some of these are normal responses to changes in the environment or circumstances of an animal, while others are beyond this "normal" range and are collectively referred to as stress. In the strict sense, stress has been defined as a state resulting from environmental conditions which threaten survival (Brett, 1958). However, it is now widely accepted that conditions which are not life-threatening but which impair feeding, growth, reproduction or other aspects of the normal performance, physiology and activity of an animal are stressful. The stimuli producing these changes are collectively referred to as stressors. Such stress may be counter-productive in the laboratory and in aquaculture it can lead to immediate or delayed mortalities and often causes poor feeding reactions for a day or so, with consequent slower growth and poorer growth efficiency.

It should be borne in mind that, because stress responses are induced in fish by changes in their environment and by netting and handling, etc., it is almost impossible to investigate the physiology of a truly unstressed fish other than by remote sensing techniques such as biotelemetry.

The adrenergic system and the hypothalamic-pituitary-interrenal axis

Animals respond to abnormal changes in their environment (stressors) by nerve-mediated motor actions and by releasing one or more hormones into the bloodstream following nervous stimulation of endocrine organs. Hormonal responses to stress in invertebrates are not well understood, but in fish and other vertebrates the hormonal actions are mediated by the adrenergic system and the hypothalamic-pituitary-interrenal axis (HPI). The approximate location of the endocrine tissues involved in the stress response in fish is shown in Fig. 1.

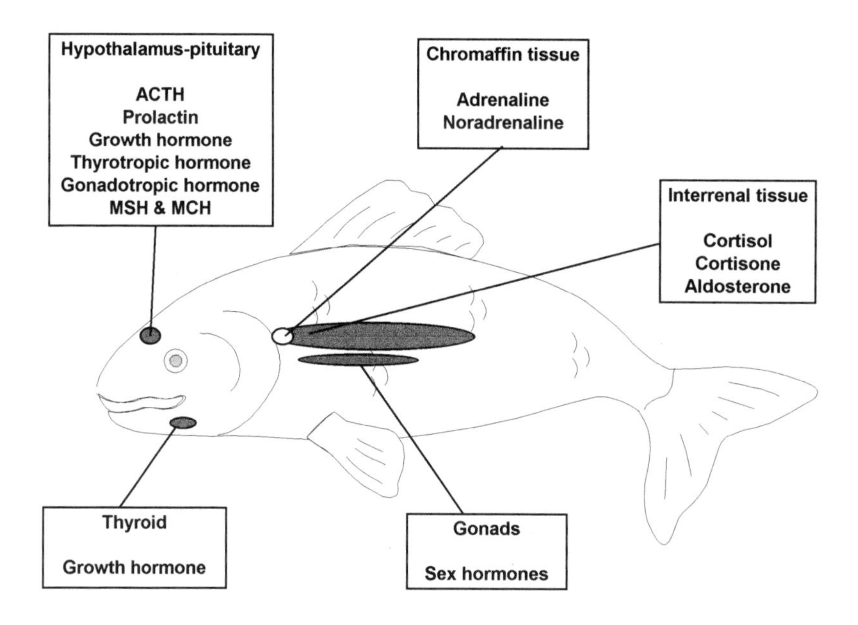

Fig. 1. The location of the principal endocrine tissues involved in stress responses in fish.

The chromaffin tissue is located in the anterior part of the kidney in fish and, following nervous stimulation, it is the source of the catecholamines adrenaline and noradrenaline (epinephrine and norepinephrine) which are released directly into the bloodstream. The hypothalamus is located in the base of the brain and the tiny pituitary gland is appended directly to it. Both areas are joined by nerve tracts running in the brain itself. A range of powerful hormones is produced in these tissues and released into the bloodstream on demand. Many of these hormones have direct metabolic effects but adrenocorticotrophic hormone, ACTH, acts on the interrenal tissue, which is also located in the kidney, to affect the synthesis and subsequent release of further powerful steroid hormones, principally cortisol. In addition to these major influences, the thyroid hormones and sex hormones may also have an influence on long-term responses to stress. It will be appreciated that the response to a stressor is not a straightforward matter but frequently involves a cascade of events, the nature of which can be visualised by reference to Fig. 2.

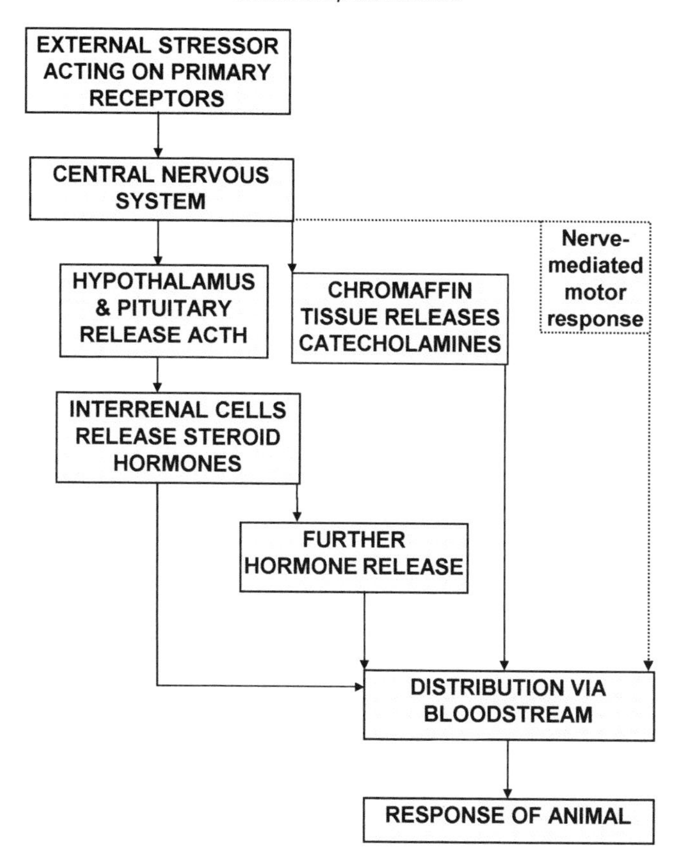

Fig. 2. Schematic diagram of stress-response sequence in fish, showing alternative control pathways.

The release of hormones from these tissues is influenced by the nature and intensity of the stressor, but may also be influenced by the age and sex of animals and the environmental temperature.

Rapid increase of circulating catecholamines tends to cause immediate reactions (the so-called fight-or-flight reaction), within seconds to hours. By contrast, release of the steroid hormones causes less immediate but usually longer-term effects, from a number of hours to several days. The response to these hormones is fairly consistent and, although their release adapts an animal to respond to an environmental change, there is strong evidence that even a minute increase in blood cortisol, when given by injection, ultimately has a deleterious effect in fish. Thus, stress can lead to increased disease susceptibility, growth depression and reproductive failure, all of which appear to be able to be caused by cortisol release, and which in some cases can be directly mimicked by direct cortisol injection or implantation (see Pickering, 1993).

The generalised stress response (GSR)

Over the last 30 years a considerable amount of work has been carried out on stress in fish (e.g. Wedemeyer, 1970; Wardle, 1972; Wardle and Kanwisher, 1974; Casillas and Smith, 1977; Soivio *et al.*, 1977; Ross and Ross, 1983; Yaron *et al.*, 1983; Pickering, 1994) and although much literature exists on the subject, only limited generalisations can be made. In recent years, however, several authors have attempted to define a general set of responses to stress which would occur in response to all stressors. In general, the GSR is characterised by an initial phase of response to the stressor, followed by a period of attempted resistance to the stressor. Depending upon the magnitude and duration of the stressor, these two phases may be followed by a period of exhaustion, in which the health or even the life of the animal can be strongly compromised. The GSR is the subject of some disagreement among researchers as to its consistency and even its existence in a sufficiently generalised sense. The reason for this is simply that the response to stressors is highly variable. The effects of stressors have been best investigated in fish and may be externally or internally evident.

External signs of stress

There are a number of characteristic external signs of stress in fish, including ataxia, obvious tachyventilation and marked colour change which can either be darkening or blanching.

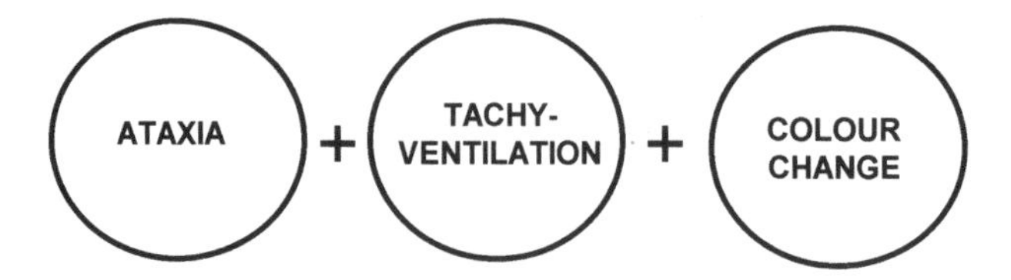

Ataxia is evident as random swimming, often at high speed and in short bursts with frequent changes of direction. This may be accompanied by body-twisting and even swimming upside-down. Brief, very high-speed movements, often in response to relatively small stimuli, are especially facilitated in fish by the giant Mauthner axons which run the full length of the spinal chord and whose stimulation causes widespread and instantaneous stimulation of motor nerves. Good husbandry and knowledge of normal stock behaviour will enable quick detection and recognition of changes in motor pattern.

Tachyventilation is evidenced by a rapid movement of the opercula in fish. Although some knowledge of the normal state must be gained before this can be judged by eye, husbandrymen or experienced research workers will have little difficulty in identifying this condition. Ventilatory and other respiratory parameters are good indicators of stress, and although the causes are not fully understood, cough rate has been found to be a better indicator than other general respiratory changes (Sprague, 1971). All fish cough at regular intervals, momentarily reversing the flow of water over the gills; this occurs about once per minute in trout and 5 times per minute in tilapia. Figure 3 demonstrates the effects of a prophylactic formalin treatment on ventilation rate and cough rate of rainbow trout, *Onchorynchus mykiss*. The massive increase in cough rate can be seen to be many times greater than the relative increase in ventilation rate, making cough rate a more useful measure of stress.

All fish have the ability to change their colour to some degree, within a certain intensity and range of the visible spectrum. They achieve this by controlling the distribution of pigment in the chromatophores of the skin both by nervous and hormonal means. The chromatophores may contain different pigments and so, in addition to lightening and darkening, animals may be able to change colour to a varying degree. The virtuosity of flatfishes is well known in this respect and some can even adapt to a chequered background. Some crustaceans have similarly versatile and frequently colourful abilities, principally under hormonal control. However, the prize for the most spectacular and legendary colour change ability goes to the cephalopod molluscs whose nervously controlled colour change is breathtaking in its speed and colour range. The problem of background colour has been recognised and experimented with by aquaculturists, but in reality it is insufficiently understood, no standard approach is taken to the problem and frequently it is simply ignored. However, because pigment distributions are controlled by both nervous and hormonal means, it is fair to assume that animals in an extreme state of pigment dispersion or concentration are stressed, at least initially. This stress may be chronic, depending upon the culture system, and some thought should probably be given to matching the background found in the natural environment, as far as possible.

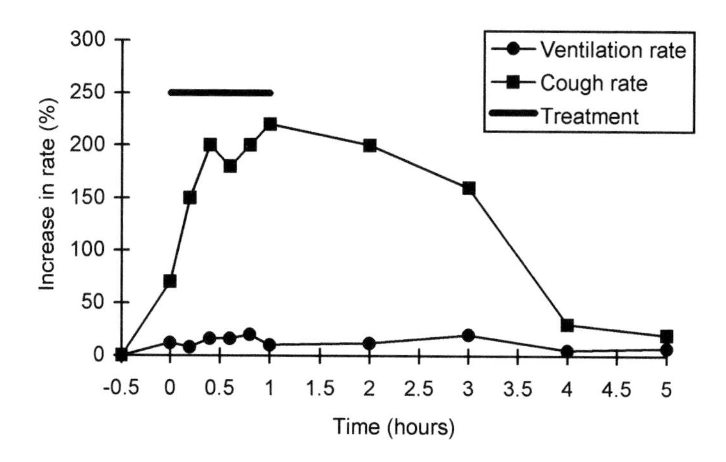

Fig. 3. The effects of a standard, prophylactic 200 ppm formalin treatment on ventilation rate and cough rate of rainbow trout, *Onchorynchus mykiss*. Based on data from Ross *et al.* (1985).

Internal signs of stress

A wide range of internal responses to stress have been described. These may be effects on the heart rate, effects on blood parameters and metabolic or biochemical effects.

Effects on heart rate

At the simplest physiological level, stress may inhibit one or more heartbeats in response to a momentary stimulus (Fig. 4) although, generally, such a small stimulus would not have a negative effect. More severe stress may, however, produce very prolonged effects which may or may not revert to normal for more than 24 hours. For example, short duration bradycardia may be replaced by tachycardia, often of long duration, and Fig. 5 shows the effects on heart rate of simple ECG electrode implantation in saithe,

Pollachius virens. Cardiac rate and ventilatory rate in fish are functionally linked and in many circumstances they are synchronised neuronally, especially when any workload is placed on the animal (Randall, 1970). As will be seen later, this link can be exploited during recovery from certain types of anaesthesia.

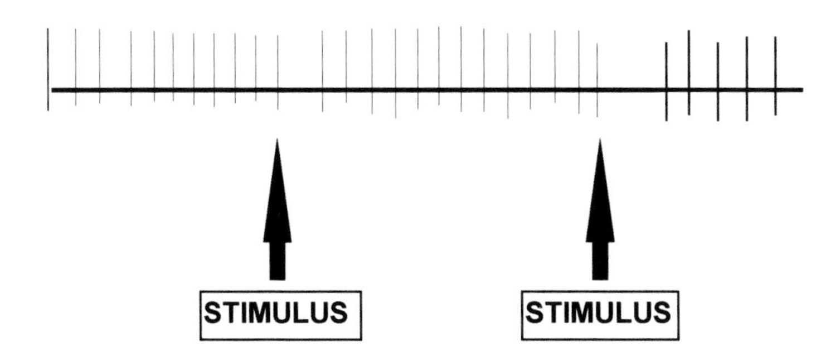

Fig. 4. The effects of stress on heart rate in saithe (*Pollachius virens*). Momentary bradycardia caused by an observer appearing at the side of the tank.

Haematological effects

There are a number of well-documented haematological effects induced by stress. These may include haemoconcentration or haemodilution, swelling of erythrocytes and increase or decrease in plasma osmolarity, chloride, sodium, potassium and other ion levels. Smith (1982) summarised the work of numerous authors to show the direction of change for a variety of physiological functions under the influence of different stressors. The relative inconsistency of these responses can be seen in Table 1. These factors not only affect circulatory physiology, but will also clearly affect osmoregulatory capability.

Hormonal effects

As already discussed, virtually all of the components of any general stress response are mediated by two groups of hormones: the catecholamines and the hormones of the HPI axis, principally cortisol.

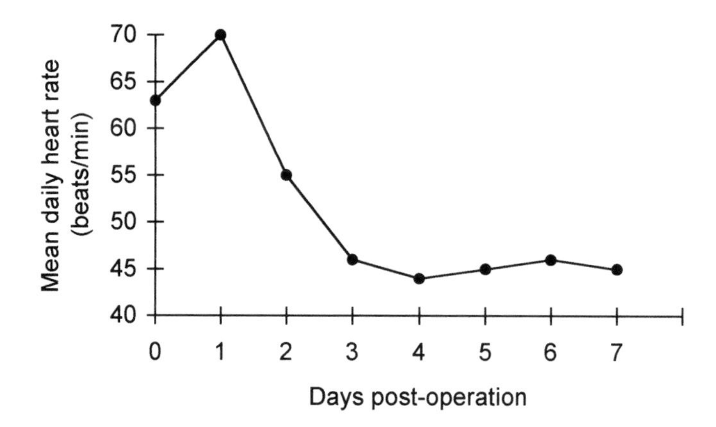

Fig. 5. The effects of stress on heart rate in saithe (*Pollachius virens*). Chronic tachycardia following anaesthesia and implantation of ECG electrodes.

Table 1. Effects of stress on blood parameters in fish. Based on data from Smith (1982). ↑ Indicates an increase and ↓ a decrease in the level of a parameter.

Parameter	Effect	Direction of change
Haematocrit (pcv)	Concentration/dilution	↑↓
Haemoglobin	Increase/decrease	↑↓
Individual erythrocyte volume	Swelling of cells	↑↓
Electrolytes	Changes in Na/K/Cl levels Changes in Mg/Ca/PO_4 levels	↑↓ ↑↓
Bicarbonate CO_2	Variable	↑↓
Glucose	Variable	↑
Lactate	Variable	
Total protein	Decreases	
Plasma concentration	Changes in osmolarity	↓
Leucocrit	Changes in cell counts	↑↓

The catecholamines epinephrine and norepinephrine, released by the chromaffin tissue of the head-kidney in response to stimulation of the sympathetic nervous system, produce a series of significant changes which are summarised in Table 2. These changes are relatively fast and may begin after less than a second and can last from many minutes up to a few hours.

Release of cortisol from the interrenal bodies in the middle region of the kidney begins in under an hour but may continue for weeks or even months (Smith, 1982). Figure 6 shows the typical rise in serum cortisol induced in response to a very simple stressor. Its effects are wide-ranging, often deleterious if prolonged and are summarised in Table 2. Its main effect appears to be an increase in catabolism and an increase in permeability of membranes to ions and its production due to stress has been well described by Wedemeyer (1969), Fagerlund and Donaldson (1970), Fryer (1975), Singly and Chavin (1975) and Yaron *et al.* (1983). Note that the increased anabolic effects are accompanied by increased catabolic effects and result in generally inhibited growth. A classical example of this is migrating Pacific salmon where cortisol levels may be elevated by up to eight times for many weeks. The net result of this is to massively alter metabolism so as to enable the energetically demanding migration and subsequent fighting and spawning activities. The process is irreversible, however, and the extensive steroid-mediated net catabolism which occurs leads to the death of all migrant Pacific salmon; none return to the sea.

Acclimation to stressors

Fish, at least, are known to be able to adapt to repeated exposure to acute stressors and usually exhibit a reduced response as the number of exposures increases. The number of mucous cells per unit area of skin of rainbow trout increased after repeated handling (Fig. 7) but then slowly returned to the original value, even though the fish continued to be handled on a daily basis (D. McCaldom and L.G. Ross, unpublished data). By contrast, the magnitude of changes in plasma glucose and lactate did not reduce in the same trial. A further excellent example of this acclimation is shown in Fig. 8, in which Pickering and co-workers used standardised cleaning actions to stress brown trout and recorded clear acclimation to the stressor as the frequency of exposure increased.

Table 2. The possible range of effects of catecholamine release and corticosteroid release during stress in fish. Data drawn from the work of numerous authors.

Agent	Effect
The catecholamines (epinephrine and/or norepinephrine)	Tachycardia and increased cardiac output Tachyventilation Vasodilation or vasoconstriction Increased blood glucose Increased blood lactate Changes in free fatty acids Increase in haematocrit Glucogenesis in liver and muscle Increased peristalsis
Steroids (Cortisol, or cortisone)	Increased protein mobilisation Increased protein synthesis Inhibition of growth Reduced utilisation of carbohydrate Increased glucose production from tissue protein Deposition of glycogen in the liver Changes in membrane permeability Increased production of and activity of Na/K-ATPase Increased leucocyte count Immunosuppression

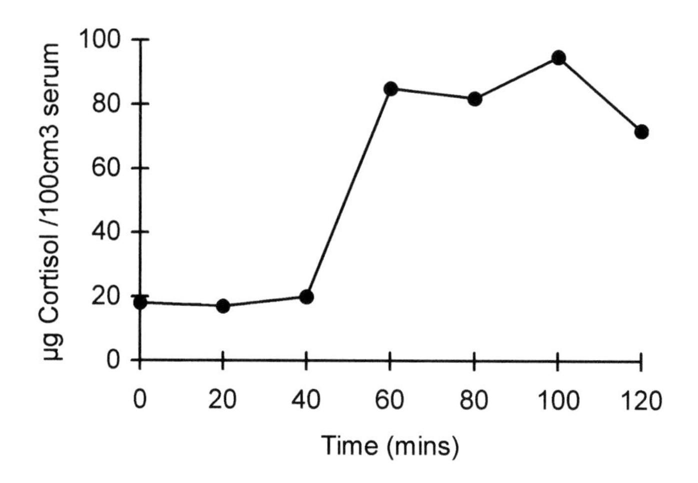

Fig. 6. Cortisol production in pink salmon, *Oncorhynchus kisutch,* stressed by stirring the aquarium water. After Wedemeyer (1969).

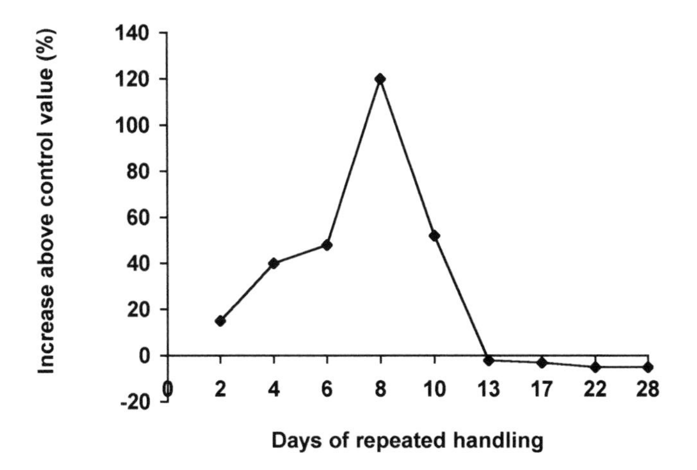

Fig. 7. Percentage change in superficial mucous cells of the shoulder region following repeated daily handling of rainbow trout, *Onchorhynchus mykiss*. Data from McCaldom (1979).

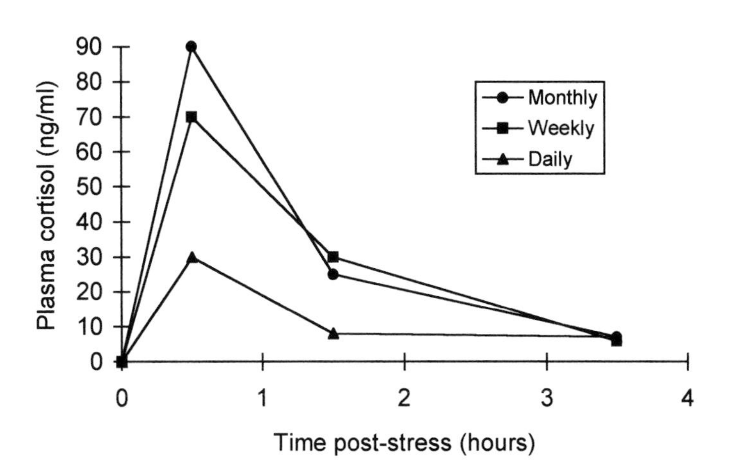

Fig. 8. Acclimation of brown trout , *Salmo trutta*, to repeated exposure to tank cleaning activities. After Pickering (1993).

Following prolonged exposure to chronic stressors, levels of plasma cortisol may return to normal levels even though the stressor is still present. However, Pickering (1993) notes that this is not a standard response and that cortisol levels may be elevated for considerable periods, ranging from weeks to months.

Stress reduction

Based upon this substantial background, it is possible to propose techniques for general stress reduction. Most recently, Pickering (1992, 1993) has summarised current thinking and has suggested a number of approaches for use in fish holding, transportation and general aquaculture. Table 3 shows a summary of practical measures which can be taken to reduce stress in any aquatic animal. Not all of these actions may be feasible in practice, although with a little thought and willingness it will be found that many can be used beneficially in a number of situations.

For the future, two routes have been proposed: (a) the breeding of "low-stress" animals, i.e. animals with a minimal response to stressors, and (b) the use of cortisol-suppressing drugs (see Pickering, 1992). Good examples of the former are found in the poultry industry and there is evidence that the magnitude of stress responses may also be heritable traits in fish. Dexamethasone, and other molecules, are cortisol receptor blockers and may have some potential for suppressing the stress response.

Stress reduction during anaesthesia

The use of anaesthesia during normally stressful procedures can alleviate, or delay, many of the normal stress reactions. Ross and Ross (1983) noted that benzocaine anaesthesia gave only a mild reduction in oxygen consumption during handling, but that this deficit had to be made up later, and strongly elevated oxygen consumption levels were recorded during recovery. Pickering and co-workers (1993) showed that the normal cortisol response may be suppressed by use of anaesthesia during handling (Fig. 9), although, again, this becomes fully manifested once the anaesthetic is withdrawn. This suppression was particularly marked with etomidate, almost certainly because etomidate is now known to inhibit cortisol synthesis.

Table 3. Summary of measures to ameliorate stress during holding, handling and transportation of aquatic animals. Adapted from recommendations intended for fish by Pickering (1993).

Suggested action	Reason	Application	Qualifications
Reduce duration of exposure to stressor	Stress response is usually proportional to duration of exposure	General	Some effects may result in long recovery times
Work at lower temperatures	Stress-induced mortality increases with water temperature	General	Not always practicable in field conditions
Prevent simultaneous stresses	Stressors may be additive or synergistic	General	Could allow time between exposure to stressors, if in a sequence
Use dilute salt solutions	Medium becomes more isotonic, reducing osmotic losses and mortality	Transportation	Use with stenohaline freshwater animals may be limited
Withdraw food: 2-3 days in cold water 12-24 hours in tropicals	Reduces oxygen requirements and fouling of medium	General Handling Transportation	Requires additional management input
Reduce stocking density or numbers handled per batch	Reduces interaction and abrasion	General Handling Transportation	May conflict with intensification of aquaculture
Provide adequate, more "natural", environment	Providing environmental "comfort" (see earlier)	General	Not always possible in some culture systems
Use mild anaesthesia or sedation	Facilitates handling, *temporarily* reduces oxygen consumption, reduces carbon dioxide and ammonia output	Handling Transportation	Anaesthetics can act as stressors

Harrell (1992) attempted to alleviate stress by using salt solution during MS222 anaesthesia, assessing stress by measuring plasma corticosteroid and chloride levels. In a comparison of 25 $mg.l^{-1}$ MS222, 25 $mg.l^{-1}$ MS222 with 10 $g.l^{-1}$ salt and salt alone, he showed that stress was least when only salt was used.

Stress induced by anaesthesia

Anaesthesia itself induces stress reactions although it is difficult to distinguish the direct effects of the anaesthetic method from those of capture or handling. It has been shown by Serfaty *et al.* (1959) and Houston *et al.* (1971) that MS222 (tricaine methane sulphonate) induction produces both tachycardia and tachyventilation. Randall and Smith (1967) demonstrated the magnitude of the cardio-ventilatory effects of MS222 anaesthesia in rainbow trout (Fig. 10) and Houston *et al.* (1971) in a comparable study showed that dorsal aortic blood pressure was similarly affected. Despite this it is clear that anaesthetics are necessary in many cases to limit the magnitude of the stress response in addition to facilitating handling.

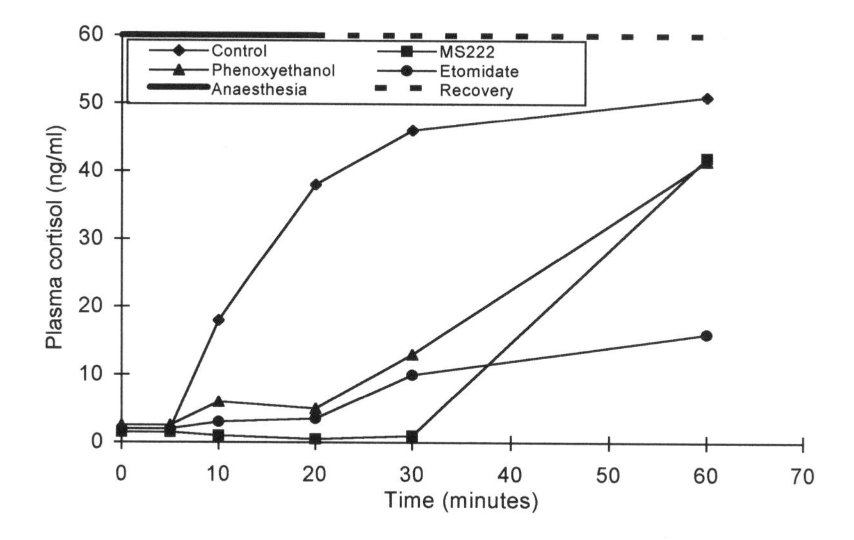

Fig. 9. The effects of anaesthesia on the cortisol response to handling in rainbow trout. After Pickering (1993).

In summary

Clearly, there may be powerful consequences of exposure to a stressor which can affect physiology, behaviour, growth and performance of cultured or experimental animals. The stress-reduction methods which have been proposed will not always be possible and may sometimes appear to be in conflict with other objectives, for example those of aquaculture. Furthermore, obviously, mild anaesthesia will not suffice in all cases. Deeper anaesthesia is required for surgery or invasive work of any kind. A range of techniques are available for the sedation or anaesthesia of aquatic animals in different circumstances and in the remainder of this book we attempt to d[illegible]e the various approaches to practical anaesthesia.

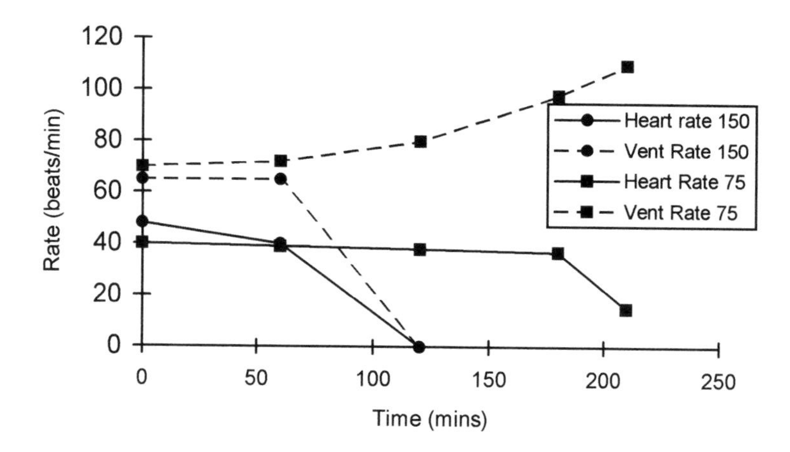

Fig. 10. Changes in heart rate, ventilation rate and ventilation amplitude at two levels of MS222 anaesthesia, 150 mg.l^{-1} and 75 mg.l^{-1}. After Randall and Smith (1967).

References

Brett. J.R. (1958). Implications and assessments of environmental stress. In: *Investigations of Fish-Power Problems* (Edited by P.A. Larkin). H.R. Macmillan Lectures in Fisheries, University of British Columbia. pp.69-83.

Casillas, E. and Smith, L.S. (1977). The effect of stress on blood coagulation and haematology in rainbow trout. *Journal of Fish Biology.* 10: 481-491.

Fagerlund, U.H.M and Donaldson, E.M. (1970). Dynamics of cortisone secretion in sockeye salmon (*Oncorhynchus nerka*) during sexual maturation and after gonadectomy. *Journal of the Fisheries Research Board of Canada.* 27: 2323-2331.

Fryer, J.N. (1975). Stress and adrenocorticosteroid dynamics in the goldfish *Carassius auratus. Canadian Journal of Zoology.* 53: 1012-1020.

Harrell, R.M. (1992). Stress mitigation by use of salt and anaesthetic for wild striped bass captured from broodstock. *Progressive Fish Culturist.* 54 (4): 228-233.

Houston, A.H., Madden, J.A., Woods, R.J. and Miles, H.M. (1971). Some physiological effects of handling and tricaine methanesulphonate anaesthetization upon the Brook Trout. *Salvelinus fontalis. Journal of the Fisheries Research Board of Canada.* 28 (5): 625-633.

McCaldom, D.M. (1979). The effect of stress on various blood and skin parameters in rainbow trout. Unpublished B.Sc. thesis, University of Stirling. 44pp.

Pickering, A.D. (1992). Rainbow trout husbandry: management of the stress response. *Aquaculture.* 100: 125-139.

Pickering, A.D. (1993). Husbandry and stress. In: *Recent Advances in Aquaculture.* Vol. 4 (Edited by J.F. Muir and R.J. Roberts). Blackwell Science, Oxford. 340pp.

Randall, D.J. (1970). The circulatory system. In: *Fish Physiology. Vol. 4. The Nervous System. Circulation and Respiration.* (Edited by W.S. Hoare and D.J. Randall). Academic Press, London.

Randall, D.J. and Smith, L.S. (1967). The effect of environmental factors on circulation and respiration in teleost fish. *Hydrobiologia.* 29: 113-124.

Ross, B. and Ross. L.G. (1983). The oxygen requirements of *Oreochromis niloticus* under adverse conditions. *Proceedings of the First International Symposium on Tilapia in Aquaculture.* Nazareth, Israel. pp.134-143.

Ross, L.G., Ward, K.M.H. and Ross, B. (1985). The effect of formalin, malachite green and suspended solids on some respiratory parameters of rainbow trout, *Salmo gairdneri. Aquaculture and Fishery Management.* 16: 129-138.

Serfaty, A., Labat, R. and Quiller, R. (1959). Les reactions cardiaques chez la carpe (*Cyprinus carpio*) au cours d'une anaesthesie prolongee. *Hydrobiologia.* 13: 134-151.

Singly, J.A. and Chavin, W. (1975). The adrenocortical-hypophysical response to saline stress in the goldfish. *Carassius auratus* L. *Comparative Biochemistry and Physiology.* 51A: 749-756.

Smith, L.S. (1982). *Introduction to Fish Physiology.* T.F.H. Publications. 352pp.

Soivio, A., Nyholm, K. and Huhti, M. (1977). Effects of anaesthesia with MS222, neutralised MS222 and benzocaine on the blood constituents of Rainbow trout. *Journal of Fish Biology.* 10: 91-101.

Sprague, J.B. (1971). Measurement of pollutant toxicity to fish. III. Sublethal effects and "safe" concentrations. *Water Research.* 5: 245-266.

Wardle, C.S. (1972). The changes in blood glucose in *Pleuronectes platessa* following capture from the wild: a stress reaction. *Journal of the Marine Biological Association of the U.K.* 52: 635-651.

Wardle, C.S. and Kanwisher, J.W. (1974). The significance of heart rate in free-swimming cod, *Gadus morhua*: some observations with ultrasonic tags. *Marine Behaviour and Physiology.* 2: 311-324.

Wedemeyer, G. (1969). Stress-induced ascorbic acid depletion and cortisol production in two salmonid fishes. *Comparative Biochemistry and Physiology*. 29: 1247-1251.

Wedemeyer, G. (1970). The role of stress in the disease resistance of fishes. *American Fisheries Society Symposium on Disease of Fish and Shellfish.* Special Publication No. 5.

Yaron, Z., Ilan, Z., Bogomolnaya, J. and Vermaak, P. (1983). Steroid hormones in two tilapia species. *O. niloticus. Proceedings of the First International Symposium on Tilapia in Aquaculture*. Nazareth, Israel.

~~~~

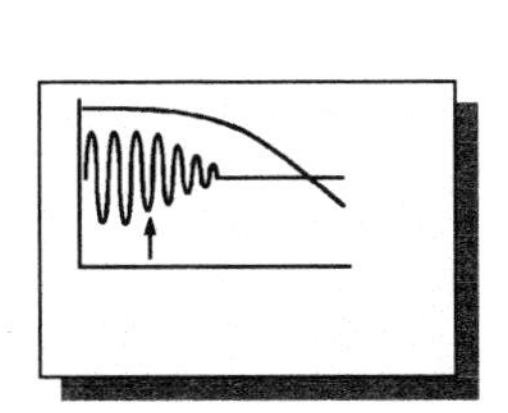
~~~~

Chapter 3

The Nature of Anaesthesia, Sedation and Analgesia

General anaesthesia and sedation

The word anaesthesia (USA=anesthesia) has a Greek derivation, meaning loss of sensation or insensibility. It can take a number of forms which require some definition. In general terms, sedation (defined as a calming effect) is a preliminary state of anaesthesia in which drowsiness is induced, with dulled sensory perception and perhaps with some analgesia (insensibility to pain), but in which there is no gross loss of sensory perception or of equilibrium. Full general anaesthesia may be defined as "a reversible, generalised loss of sensory perception accompanied by a sleep-like state induced by drugs or by physical means" (Heavner, 1981). Green (1979) notes that "general anaesthesia involves a state of general depression of the CNS involving hypnosis, analgesia, suppression of reflex activity and relaxation of voluntary muscle". The term **narcosis** is also frequently used in discussing anaesthesia and anaesthetic agents. A narcotic is defined as a "sleep inducing agent whose effect varies according to the strength and amount of the narcotic agent administered". In animal work, anaesthesia always assumes that recovery will, or could, occur whereas narcosis may be used where recovery is not anticipated. There is clearly a great deal of overlap in the use of these terms and narcosis and anaesthesia are essentially synonymous.

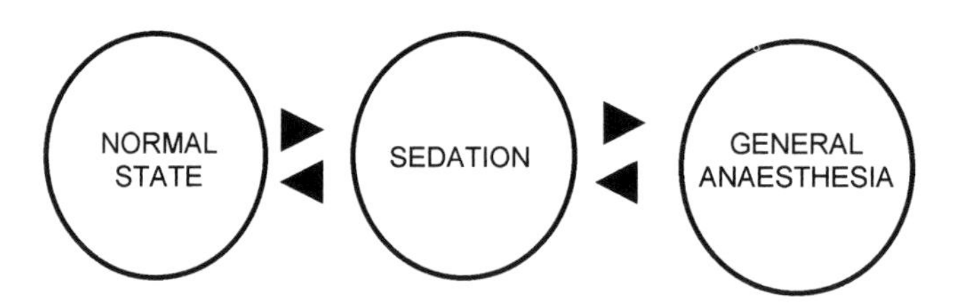

General anaesthesia and sedation can be produced by a variety of techniques including physical, chemical and psychological methods, which are summarised in Table 4. Most of these techniques apply to aquatic animals, although some, notably acupuncture, pressure, rectal drug administration and suggestion, almost certainly do not.

Psychological methods operate by drawing the subject's attention to something other than the noxious or aversive stimulus and it is postulated by some that acupuncture operates in a similar way. There is some support for the notion that acupuncture produces a "gating" effect in the central nervous system, either physically or by suggestion or a combination of both. Although some terrestrial animals seem to respond to this form of anaesthesia, there have been no attempts to use it in aquatic animals. Surprisingly, a form of hypnosis, known as catatonia, seems to be possible in large crustaceans but as far as is known this has no practical use in aquaculture or biological studies.

Table 4. Methods used to produce general or local anaesthesia.
Adapted from Heavner (1981).
Techniques in italics have not been shown to be effective in aquatic animals.

Method	Technique
Physical	Hypothermia
	Electrical stimulation
	Acupuncture
	Pressure
Chemical (drugs)	Inhalation
	Parenteral: Intravenous Intraperitoneal Intramuscular
	Oral
	Rectal
Psychological	*Suggestion/distraction*
	Hypnosis

Local anaesthesia

Local anaesthesia is "a reversible loss of sensation in a discrete region of the body" (Heavner, 1981). Local anaesthetic drugs are administered directly to the area where they are needed to act and they do not need to be absorbed into the bloodstream to be effective. They are given either by injection or by applying the drug to the surface, usually the skin, a procedure technically known as topical block. Local anaesthetics produce their effect by blocking propagation of action potentials in receptors and nerves at, and near to, the selected site, effectively preventing transmission of any painful stimulus to the central nervous system. In addition to using drugs, local anaesthesia can also be produced by the use of locally-applied hypothermia, electric current, acupuncture, application of pressure and even hypnosis (see Table 5) although, clearly, many of these methods would not be effective in aquatic animals.

Defining pain

As already noted, pain control may be required in procedures with aquatic animals. Despite considerable research, our understanding of the mechanisms of pain production and reception is by no means complete (Lineberry, 1981). The problem is partly that pain is a complex experience comprising both an unpleasant sensory reaction to a stimulus and an emotional component. The painful stimulus may also produce some tissue damage, in which case it is described as noxious, although there is not strict adherence to the use of these terms. Melzack (1973) considers that a painful stimulus must also be accompanied by an aversive response, usually behavioural, which may be expected to end the unpleasant experience or at least be intended to do so. Pain may be produced by mechanical, thermal, electrical or chemical stimulation, or a combination of some of these. Because of this it is difficult to easily define the stimulus energy required to elicit pain. The problem is much greater in animals than in humans as it is, of course, not possible to converse with animals directly and hence to build up an objective, or even subjective, description of painful events which help the observer to understand the process.

There is much discussion as to whether some animals perceive pain in the same way as humans, or even at all. There is little doubt that painful stimuli are perceived by most, if not all, animals. Even some relatively very simple invertebrates possess definable pain receptors, nociceptors, which are well known to nerve physiologists. This is, however, somewhat philosophical, frequently involves relatively subjective anthropomorphic interpretation and bias on both sides of the argument and opposing views can be taken depending upon interpretations of the available data. Because of these qualifications, in animal research pain is usually described in an operational sense whereby a stimulus is assumed painful if it normally produces pain in humans, can lead to tissue damage and if an escape behaviour is elicited. Most painful stimuli will, or could, produce all three effects.

Analgesia

Analgesia is the relief from pain. It is a block of pain perception with or without the retention of other sensory abilities and with or without continued control of motor function. It typically does not result in loss of equilibrium. In animals it is reflected by a reduced response to noxious stimuli. Again, because of the nature and definition of pain itself, and the inability to converse, analgesia in animals may be difficult to judge. It is further complicated by the fact that the response to normally noxious stimuli can change and new responses can be learned. A classical example is that of Pavlov's dogs (Pavlov, 1927), who learned to associate a food reward with an electrical stimulation of the paw. The normal violent reaction to the painful electrical stimulus was soon replaced by tail-wagging and salivation. Research into pain relief in animals is used as a means of studying pain alleviation in humans. It is however, difficult on technical grounds and it has become increasingly difficult to justify the use of animal models on ethical grounds.

Overall, pain relief or analgesia in aquatic animals may be judged to have occurred when the subject no longer responds, or has a markedly reduced response, to normally noxious stimuli following sedation or anaesthesia.

The mechanism of anaesthesia

General anaesthetic procedures usually act by widespread depression of the central nervous system produced by an action on nerve axons, transmitter release or membrane excitability (Fig. 11). The only unifying principle of general anaesthesia is that anaesthetics interact with membrane components and no single cellular mechanism appears to be able to explain their widespread effects in the central nervous system (Winlow *et al.*, 1992).

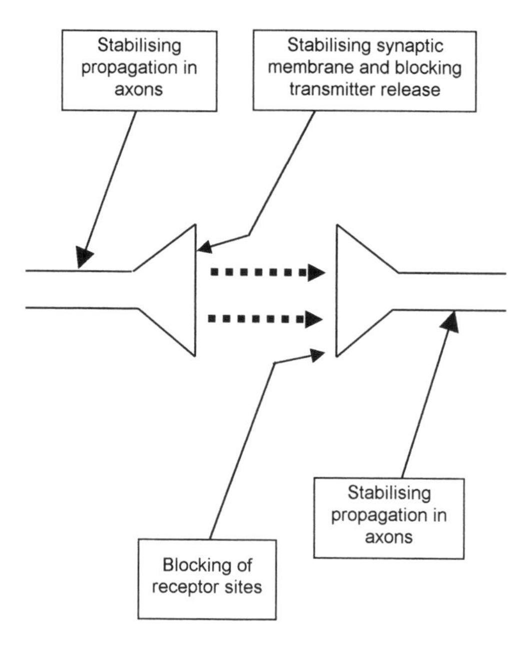

Fig. 11. Some probable mechanisms of action of anaesthetic agents.

It should be noted that little is known about the precise mode of action in invertebrates and fish, although with some drugs there appears to be an inverse relationship between the dose level required to induce a given depth of anaesthesia and the animal's evolutionary status. Consequently a fish would require a larger dose of a drug than a mammal to produce a given effect. It has been suggested that this phenomenon may be due to an increasing presence of active sites in higher vertebrates for any particular molecular form.

The stages of anaesthesia

When induction is slow a series of stages of anaesthesia can be observed. This occurs in most animals but was first described for fish by McFarland (1959). The basis of his descriptive scheme is summarised in Table 5 and it can be seen that an anaesthetic substance can produce sedation, surgical anaesthesia or death, depending on the combination of the dose level and the length of exposure.

Table 5. Stages of anaesthesia in fish. After McFarland (1959).

Stage	Plane	Description	Physiological and behavioural signs
I	1	Light sedation	Responsive to stimuli but motion reduced, ventilation decreased
	2	Deep sedation	As above, some analgesia, only receptive to gross stimulation
II	1	Light anaesthesia	Partial loss of equilibrium. Good analgesia
	2	Deeper anaesthesia	Total loss of muscle tone, total loss of equilibrium, ventilation almost absent
III		Surgical anaesthesia	As above: total loss of reaction to even massive stimulation
IV		Medullary collapse	Ventilation ceases, cardiac arrest, eventual death. Overdose

In practice, it is often found that a species may not conform clearly to every aspect of McFarland's description, although there is sufficient general agreement for this to serve as a good preliminary basis.

For practical purposes, anaesthesia reduces to three obvious phases: **induction, maintenance** and **recovery.** These principal phases are referred to from time to time in the literature, but unfortunately with no consistency. Each of these varies in duration according to drug or method used, species and conditions.

During induction the animal is exposed to the anaesthetic agent in order to achieve the desired stage. Induction is often accompanied by hyperactivity, usually a response of only a few seconds to the slightly irritant properties of the drug. In general, induction should be rapid and without marked hyperactivity, although there is usually some. The animal will exhibit a succession of the signs noted in Table 6, notably ataxia and loss of the righting reflex, eventually passing into surgical anaesthesia with no reaction to any stimuli.

Maintenance involves extending the achieved stage in a stable manner without detriment to the health of the animal. It should be uneventful and will most probably be effective on a reduced drug dose. The environment of the anaesthetised animal needs to be controlled and physiological needs such as oxygen supply and removing waste gases and metabolites must be attended to. It is usual to monitor the condition of animals during maintenance. This is achieved either visually or by measurement of ventilation rate or examination of ECG. The latter will clearly not be necessary, or even possible, where batches of animals are being handled. In research, where individual ECG monitoring is undertaken, arrhythmias should be avoided and excessive bradycardia or cardiac arrest requires immediate action to recover the animal. Over-exposure may be accompanied by "flaring" of the opercula, which may be repeated several times (Klontz and Smith, 1968). Immediate steps should then be taken to recover the animal.

The recovery phase involves withdrawal of the anaesthetic agent and return to a normal state. Initial recovery may take anything from a few seconds to a few minutes, but in general it should be quick and without altered behaviour or other side effects, although there will usually be some muscle trembling. The animal will attempt to right itself and will begin to respond to noise or other sensory stimuli. The time for a full recovery can be from minutes to days and will depend on the species and drugs used. Sympathetic treatment, handling and general good care are essential during this period to avoid post-operative death.

Dose, exposure time and effect achieved

As already described, the actual anaesthetic effect achieved will depend upon the dose or level of agent used and the length of exposure to that agent. If the agent is sufficiently strong, the animal will rapidly pass through all of the possible stages of anaesthesia. At lower concentrations a condition may be achieved where the clearance rate equals the rate of uptake and then an approximately steady state may be maintained. Understanding this relationship between dose, exposure time and anaesthetic stage achieved is very important to ensure good control of a procedure. The degree and nature of analgesia achieved and the ease of recovery is also extremely important. Unfortunately, proper descriptions of all of these features are rarely, if ever, available for most agents.

An example of the stages of anaesthesia achieved or passed through at different drug dose levels is shown in Fig. 12, in which the time to achieve certain observable criteria is recorded. The authors considered that there was a good safety margin at the lower doses and showed a clear potentiating effect of a higher temperature in this case. The recovery times are rapid at the two lowest dose levels but become more prolonged as dose level increases (Fig. 13).

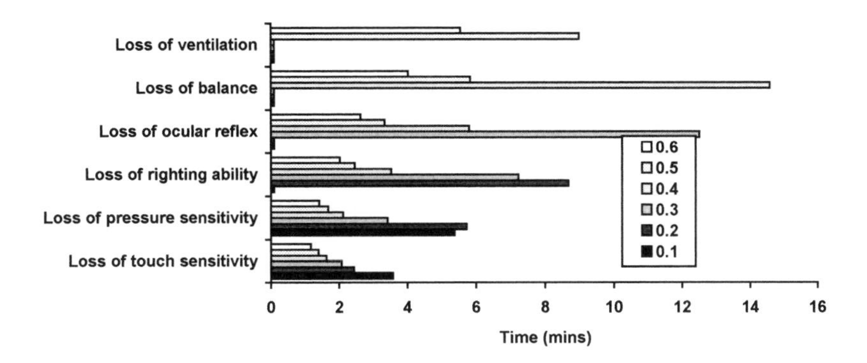

Fig. 12. Time to induction in carp (*Cyprinus carpio*) using 2-phenoxyethanol at 10°C. Dose levels are in $ml.l^{-1}$. Based on data from Josa *et al.* (1992).

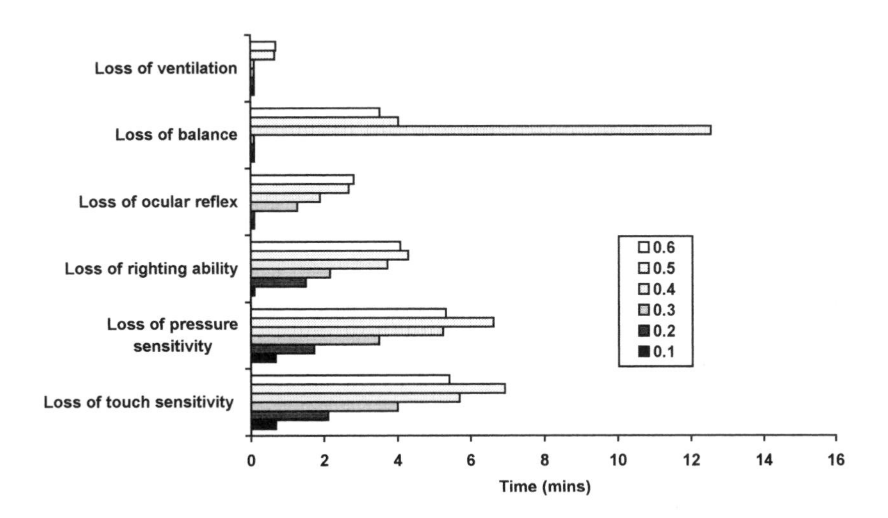

Fig. 13. Time for recovery of carp (*Cyprinus carpio*) following anaesthesia using 2-phenoxyethanol at 10°C and 20°C. Dose levels are in $ml.l^{-1}$. Based on data from Josa *et al.* (1992).

These induction and recovery effects may not be evident in the same way with all agents used but, in general, they show what may be expected.

Assessment of analgesia is very important but is rarely made. In most cases analgesia has been assumed to be effective because the animal is immobilised. This may be a reasonable assumption with some classes of anaesthetic drug, but cannot be relied upon in all cases. Robinson (1984, unpublished data) showed that, following immobilisation using electricity, there was good analgesia in rainbow trout, *Oncorhynchus mykiss,* but incomplete analgesia in tilapia, *Oreochromis niloticus,* which still responded to pressure and pain stimuli. Some assessment of analgesia should always be made when investigating new agents for aquatic animals.

Euthanasia

In some cases it will be necessary to kill animals humanely, either to terminate an experimental procedure, harvest tissues, cull surplus animals or to alleviate distress. This procedure, known as euthanasia, should be carried out with the minimum of physical and mental suffering and so it is best to prepare for this contingency in advance of carrying out any work which may possibly require it. This will involve arranging facilities in advance, as well as ensuring a minimum of disturbance to the subjects, considering safety of personnel and using a method in which you have confidence, i.e. one which is going to work quickly and reliably.

In the UK, legislation provides for two possible approaches to euthanasia, either physical or chemical. Physical methods involve either dislocation of the neck, or a sharp blow to the head. The latter technique is most frequently used and is very effective, while the former may not be possible with some species. Chemical methods involve the use of anaesthetic overdose, administered either by injection or immersion. Pithing, in which the brain and anterior spinal cord are destroyed to absolutely ensure death before disposal is effected, should follow all these methods. Pithing is carried out with a scalpel or sharp blade and a seeker or long needle. In fish, pithing would typically involve cutting through the spinal column just behind the head and then passing the long needle or seeker into the spinal cord to sever connections, followed by passing the needle into the brain cavity and destroying connections by a number of lateral and vertical movements. Clearly, the technique should be adequately practised on cadavers, with guidance and supervision, before use. Once mastered, the procedure is rapid and reliably ensures death.

In summary

A basic understanding of the nature of anaesthesia and how it immobilises and aids in pain suppression is essential to enable users to understand and control the process. This background is also important in allowing workers to judge induction, the depth of anaesthesia and of analgesia, and to be able to adapt the basic scheme to new species and circumstances. It is important to use sufficient sedation for the purpose, while at the same time avoiding overdosing, which is commonly caused by practitioners. There are many disasters, unreported of course, which occur as workers rush headlong into untested methods. We consequently urge extreme caution and judicious pre-experimentation by all those who intend to carry out animal sedation or anaesthesia, of any kind. Anaesthesia and sedation are just part of the aquatic biologists' toolbox, but if mismanaged they can ruin everything.

References

Green, C.J. (1979). *Animal Anaesthesia*. Laboratory Animal Handbooks. No. 8. London. 300pp.

Heavner, J.E. (1981). Animal models and methods in anaesthesia research. In: *Methods in Animal Experimentation*. Vol. 6 (Edited by William I. Gay). Academic Press, New York. 400pp.

Josa, A., Espinosa, E., Cruz, J.I., Gil, L., Falceto, M.V. and Lozano, R. (1992). Use of 2-phenoxyethanol as an anaesthetic agent in goldfish (*Cyprinus carpio). Research in Veterinary Science*. 53: 139.

Klontz, G.W. and Smith. L.S. (1968). Fish as biological research subjects. In*: Methods in Animal Experimentation*. Vol. 3 (Edited by William I. Gay). Academic Press, New York. 469pp.

Lineberry, C.G. (1981). Laboratory animals in pain research. In*: Methods in Animal Experimentation*. Vol. 6 (Edited by William I. Gay). Academic Press, New York. 400pp.

McFarland, W.N. (1959). A study of the effects of anaesthetics on the behaviour and physiology of fishes. *Publications of the Institute of Marine Science*. 6: 22-55.

Melzack. R. (1973). *The Puzzle of Pain*. Basic Books, New York.

Pavlov, I.P. (1927). *Conditioned Reflexes*. Milford Press.

Robinson, E. (1984). A study of the use of alternating current for electroanaesthesia in *Salmo gairdneri* and *Oreochromis niloticus*. B.Sc. thesis. University of Stirling. 26pp.

Winlow, W., Yar, T., Spencer, G., Girdlestone, D. and Hancox, J. (1992). Differential effects of general anaesthetics on identified molluscan neurons *in situ* and in culture. *General Pharmacology*. 23 (6): 985-992.

~~~~

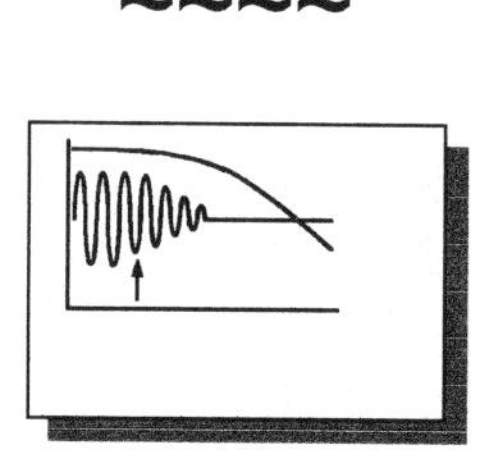
~~~~

Chapter 4

The Features of Anaesthetic Agents, and their Legal Use

Introduction

Broadly, there are three main approaches to the sedation and anaesthesia of fish, namely the use of drugs and gases, induction of hypothermia or exposure to an electric current. Certain of the features required of an anaesthetic agent are common to all methods in current use, although in some cases special considerations or specific safeguards will be needed. It is unlikely that any agent, either existing or as yet undiscovered, will be able to fulfil all of the ideal pre-requisites.

Desirable features of an anaesthetic agent

The ideal characteristics for anaesthetic agents have been remarked upon by numerous authors, most usually when considering chemical methods (Marking and Meyer, 1985). In general, anaesthesia or sedation should be induced rapidly and with minimum accompanying hyperactivity or other stress. The agent or method should be easy to administer, involving no complex procedures. It should be easy to maintain the animals in the chosen state, should provide proper immobilisation and effective analgesia. Recovery should be rapid and uneventful without prolonged ataxia or other undesirable features.

Where a sedative or anaesthetic drug or gas is used it is preferable that it should be effective at low doses and that the toxic dose should considerably exceed the effective dose, providing a wide safety margin. A predictable level of analgesia should be provided. It should also be safe to operators at the dose levels used and should not be an irritant or carcinogen. It is also desirable that the chemicals do not produce hyperactivity during induction of the subject; some materials are much more prone to inducing this effect than others. In addition, the substance should be easily soluble in water or in a water-soluble solvent that can be used as a vehicle. The drug and its solvent should be easy to obtain in bulk and should be relatively inexpensive. In aquaculture and fisheries work, large numbers of animals are frequently handled and it is often necessary to dispense large quantities of drugs. The drug should, consequently, remain effective in working solution for more than 24 hours or so and should also be chemically stable over a reasonable period of time in storage. Where a drug is unstable or degraded over a short period cost becomes more important.

When fish are immobilised by lowering of temperature, the safety margin is frequently quite small and deaths occur if temperature is lowered too far, or too quickly. Thus, the main consideration in hypothermia is that the rate of cooling can be carefully controlled to avoid mortalities and that the required reduced temperature can be easily maintained.

Where immobilisation using electricity is used, it is important that the side effects do not exceed those encountered using other methods and that the electrical stimulation does not induce violent motor responses which can disfigure, or even kill. The safety margin to the animal must be well understood. Overall, operator safety is of paramount importance, although it is very difficult to achieve in practice.

These preferred features of anaesthetic agents are summarised in Fig. 14.

Toxicity and margin of safety

As already noted, it is preferable that a sedative or anaesthetic drug or gas should be effective at low doses and that the toxic dose should considerably exceed the effective dose, thereby providing a wide safety margin. The effective dose and toxic doses are linked in much of the literature, either qualitatively in the use of the term "margin of safety" or quantitatively through the "therapeutic index". The therapeutic index is the ratio of the effective safe dose to the dose which first produces undesirable side effects (Loomis, 1968), approximated by 99% survival. Although the therapeutic index will not be available for all drugs and species combinations, it is probably the most meaningful practical index.

An alternative way to express this margin of safety is the LC_{50}, the concentration of a substance which will kill half of the sample group within a given period of time. The LC_{50} may be available for some drugs, but is arguably of less utility for anaesthetics than the therapeutic index because of the lengths of exposure most frequently used in toxicity testing (24 h, 96 h, 144 h, etc.). Many aquatic animals will be exposed to the working concentration of water-soluble anaesthetics for only a relatively brief time, usually considerably less than 30 minutes, making the LC_{50} difficult to interpret.

GENERAL FEATURES OF AN ANAESTHETIC AGENT

- Easy to administer
- Effective at low dose or exposure
- Provides sedation or anaesthesia predictably
- Provides good analgesia
- Induces desired state quickly
- Easy to maintain the desired state
- Easy to reverse the process
- Uneventful rapid recovery
- Wide safety margin
- Low cost

WHEN USING DRUGS

- The substance should be easily soluble in water or in a water-soluble solvent
- Drugs or gases should not produce hyperactivity during induction
- Should not induce a strong stress response
- The drugs should be easy to obtain in bulk.
- Drugs should be stable
- Drugs should be safe to operators and non-carcinogenic
- Leaves negligible tissue residues after a short withdrawal time
- Environmentally harmless or rapidly biodegradable

WHEN USING HYPOTHERMIA

- Required rate of cooling is known and controllable
- Maintenance of desired temperature
- Mortalities low to absent

WHEN USING ELECTROANAESTHESIA

- Safety to operators
- Safety to animal subjects
- High safety margin
- Side effects should not be significantly different from other methods
- Any necessary licensing is arranged

Fig. 14. The desirable features of anaesthetic agents.

Additives

A number of proprietary drugs used in anaesthesia and sedation contain additional materials. These are usually antioxidants designed to extend shelf life, or surfactants intended to improve flow characteristics and injection properties. These materials may have undesirable, or unknown, side effects in aquatic animals and where a new preparation is to be evaluated it is essential to have full data on the formulation, if available.

Anaesthetics and legislation

In recent years, the legislation covering procedures and materials described in this book has been steadily improved and extended, to the general benefit of animal welfare, the user community and the consumer of aquatic products. Although varying widely world-wide, much of the legislation has numerous common features, especially in Europe, North America, Australia and New Zealand. All users of the techniques described here must become aware of the relevant legislation for the country in which they operate. Because of this, the disclaimer already included in the preface is repeated here for emphasis:

"The authors and the publisher accept no responsibility for losses of animals, losses of experimental or other data, harm occurring to individuals or any other matter which may arise caused by application of the data contained herein or which may arise as a result of non-compliance with any relevant legislation."

Table 6. Summarises the broad legislative objectives relevant to pharmaceuticals in the United States, and to this must be added electrical safety. The legislative issues then fall into the following categories:

- Safe operator practice for users and safe storage of drugs and chemicals
- Safe use of drugs with animals intended for the human food chain
- Potential release of drugs into the environment
- Animal welfare and experimentation
- Safety legislation concerning electric fishing and similar electrical apparatus

Table 6. Legislative targets for licensing of pharmaceuticals in the United States. After Beleau (1992).

Point	Objective	Protected group
1	The drug must be shown to be safe to the treated animals	Cultured stock
2	The drug must pose no hazard to the user	Human users (e.g. the fish farmer)
3	No harmful residues must be left which could contaminate the human or animal food supply	Food chain; humans and other food animals
4	The material must not contaminate the environment either in its use or its manufacture	Environment; all wildlife, all vegetation, humans
5	The product must be effective for its intended purpose	Cultured stock

Safe operator practice for users and safe storage of drugs and chemicals

Use of anaesthetic agents, particularly chemicals, in the aquaculture industry is subject to a further set of constraints relating to safety to the operator. In the UK, use of chemicals in places of work and in research laboratories is regulated by a Health and Safety Executive (HSE), who operate under overarching legislation on the Control of Substances Hazardous to Health (COSSH). This requires that all users of substances, and their supervisors and co-workers, understand the hazards and risks associated with their use, that drugs and chemicals are stored and handled correctly and that proper procedures are in place for their disposal. A documented risk assessment must be carried out covering all stages from purchase to disposal and users must be formally advised of all potential hazards and be given practical instruction where needed. Any necessary protective clothing, gloves, goggles, etc., must be supplied and must be used. Similar rules exist in other countries and steps must be taken to ensure compliance, whether in the laboratory or on the farm.

Most substances used routinely in anaesthesia of aquatic animals require relatively simple storage and pose no excessive dangers to operators. Clearly, special arrangements will be needed for safe storage of gases, regulated drugs such as barbiturates, and volatile solvents such as the alcohols.

Food chain safety

Anaesthetic agents can be licensed for use in food animals in two ways: either by extending the licensed use of material already approved for use with a terrestrial farmed animal, or by a full drug development programme designed to meet current legislation. Satisfying this objective requires a wide range of inputs from the drug company, research scientists, national agencies and the farming and feed industries. The aquaculture industry is a large one world-wide, but is still relatively small compared to other animal production industries or to the human medicine industry. For this reason, drug companies may have a conflict between supporting the costs of fully licensing a new material against the expected financial returns. The time required is long and the costs are high. It is not surprising, therefore, to find that the range of anaesthetics currently approved for food use in the USA and UK is very small indeed (Table 7).

Table 7. Anaesthetic drugs approved for use in aquaculture in the USA and the UK.

Country	**Drug**	**Withdrawal time**
UK	MS222	21 days
	Carbon dioxide *	
USA	MS222	21 days
	Carbon dioxide **	–
	Quinaldine	

* Used in bleeding of salmonids. May not be true anaesthesia.

** Approved by exemption (Beleau, 1992).

As noted by Alderman (1988), the detection limit for drugs in fish and shrimp tissue is improving continuously and hence there is a real problem in defining what the acceptable tissue limits are and consequently, of course, the appropriate withdrawal times.

Environmental safety

The use of chemicals in aquaculture and fisheries and their release into the environment is of increasing concern. Legislation already exists covering a wide range of on-farm compounds. For example, in the UK, there are clear definitions for a range of materials, including organophosphates, organo-tin compounds, organo-silicon compounds, toxic metals, antimicrobials, biocides, substances affecting taste or odour, ammonia and nitrates, both in terms of discharge limits and groundwater standards. The current concerns and developments in risk assessment procedures for the management of such materials are discussed by Redshaw (1995). Although no specific UK legislation currently exists regulating discharge of anaesthetic agents, a survey by SOAFD (1992) estimated that 2 tonnes of anaesthetic agents were used in 1992 in Scottish aquaculture alone. In real terms this is a relatively small quantity and, coupled with the wide distribution of the material and the relatively low toxicity of working strengths of anaesthetic, the environmental hazards they pose are probably very low in comparison to most of the other materials mentioned above.

Animal welfare and experimentation

A common approach to regulation is now taken within the countries of the European Union and a largely similar system exists in many other countries. In the UK, approval must be sought for any procedure to be used from the Home Office Inspectorate who control all such aspects of research with animals under the auspices of the Animals (Scientific Procedures) Act (1987). It is increasingly the case that every procedure to which an experimental animal will be subjected requires approval and this now extends to anaesthesia which, as a routine preparatory procedure, was formerly excluded. Anaesthesia of aquatic animals may be used in research or in aquaculture and the choice of approach, and of drugs, for use in research is principally left in the hands of the individual who will use the most appropriate approved methods for the objectives of the work. To gain approval, the method of handling, technique for exposure to the agent, and expected nature of recovery must be formally documented, along with written contingency actions to be taken in the event of excessive pain or accidental non-recovery. Furthermore, only licensed persons may use these techniques, although they may be delegated to appropriately trained support staff under supervision in certain circumstances.

There are many disasters, unreported of course, which occur as workers rush headlong into untested methods. We consequently urge extreme caution and judicious pre-experimentation by all those who intend to carry out animal sedation or anaesthesia, of any kind and full compliance with any legislation.

Safety legislation concerning electric fishing and similar electrical apparatus

Electricity and water do not mix safely and electroanaesthesia poses special problems. In general, good construction standards, multiple electrical safety features and strict codes of operation are required to approach a sufficient degree of safety, although there can be no absolute protection against deliberate or accidental immersion in the electrical field. The nature of operator accidents ranges from mild shock to death. There are no official guidelines on this for anaesthesia, although a number of useful articles can be found in Cowx (1990). The guidelines for electric fishing equipment and practice, while useful, will be found to require modification for anaesthesia.

In summary

The desirable features of an anaesthetic agent are a wish list, a largely unattainable set of ideal parameters. However, many drugs do realise some of these features and tend to be those in wider use. The additional legislative matters affecting methodology, drug choice and safety are extremely important and are growing. Some indications of the factors which users will need to address are given here, using UK regulations as an example. Users are urged to adhere to all necessary regulations for the location in which they work.

References

Alderman, D. (1988). Fisheries chemotherapy: A review. In: *Recent Advances in Aquaculture*. Vol. 3. (Edited by J.F. Muir and R.J. Roberts). Croom Helm, London. 420pp.

Beleau. M.H. (1992). Drug development for aquaculture applications. In: *The Care and Use of Amphibians, Reptiles and Fish in Research. Proceedings of a SCAW/LSUSVM Sponsored Conference*. (Edited by Dorcas O. Schaeffer, Kevin M. Kleinow and Lee Krulisch). Scientists Center for Animal Welfare. 196pp.

Cowx, I.G. (1990). *Developments in Electric Fishing*. Fishing News Books, Oxford. 358pp.

Loomis, T.A. (1968). *Essentials of Toxicology*. Lea and Febiger, Philadelphia.

Marking, L.L. and Meyer, F.P. (1985). Are better anaesthetics needed in fisheries? *Fisheries*. 10: 2-5.

Redshaw, C.J. (1995). Ecotoxological risk assessment of chemicals used in aquaculture: a regulatory viewpoint. *Aquaculture Research*. 26: 629-637.

SOAFD (1992). Marine pollution monitoring management group. Final report of the MPMMG subgroup on marine fish farming. Scottish Fisheries Working Paper No. 3/92, February 1992.

~~~~

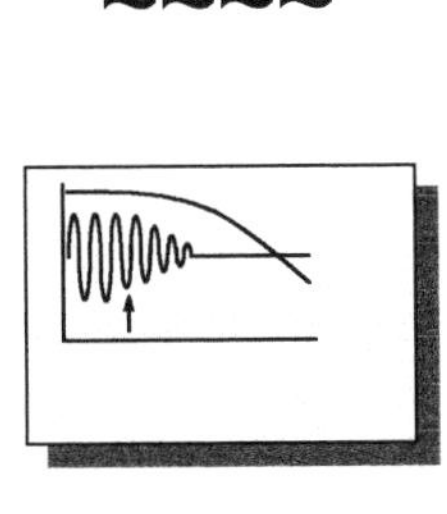
~~~~

Chapter 5

Factors Affecting the Response of Aquatic Ectotherms to Anaesthesia

Introduction

In common with anaesthesia of other animals, there is a series of factors which can alter or mediate the efficacy of anaesthetic processes in fish. These can broadly be divided into biological and environmental factors and they are summarised in Table 8.

Table 8. Factors affecting the efficacy of anaesthetics for fish.

Factor group	Factor	Probable mechanism
Biological	Species	➢ Differences in body design and habit
		➢ Gill area to body weight ratio
	Strain or genetic variance	➢ Physiological variability
	Size and/or weight	➢ Differences in enzymes?
	Sex and sexual maturity	➢ Change in metabolic rate
		➢ Lipid content?
	Lipid content	➢ Especially in gonads
		➢ Lipophilic drugs?
	Body condition	➢ Oily fish or older specimens
	Disease status	➢ Exhausted animals, post-spawners
	Stress	➢ Weakened or exhausted animals
Environmental	Temperature	➢ Q_{10} as in all ectotherms (poikliotherms)
	pH	➢ pKa effects and ionisation of molecules
	Salinity	➢ Buffering effects
	Mineral content of environment	➢ Calcium antagonism

Biotic factors

There are approaching 30,000 species of fish and, unlike other distinct vertebrate taxa, they are very diverse both in body shape and detailed design. A 100 g trout, for example, has a very different body shape and a greater gill area than an eel of similar weight. Consequently, they are unlikely to respond similarly to a given anaesthetic treatment. Often the rate at which anaesthetic drugs become effective can be related to gill area to body weight ratios and these can vary considerably between species. Indeed in some species, for example certain air breathers, the gills are reduced to a single arch used only for osmoregulation and excretion. Aquatic species with different habits have intrinsically different metabolic rates. For example, the resting respiratory rate of very active species, e.g. a tuna, may be many times greater than that of a sedentary species, e.g. the monkfish, *Lophius* sp. From knowledge of body structure and habit it is usually possible to predict how a given species may respond.

Differences in response to various drugs are well known in higher vertebrates, including different strains of mice, rats, rabbits, pigs and dogs (Green, 1979). Such differences have not been described in aquatic animals, but there is every probability that they exist and this should be borne in mind.

It has been noted by some authors that, within a species, there is a direct relationship between drug dose and size (Houston and Woods, 1972; Huish, 1972). This empirical general observation would be easy to understand if metabolic rates per unit weight increased with body weight, whereas in fact there is an inverse relationship between metabolic rate and body weight. In practice, large active fish within a group often succumb to chemical anaesthesia before their smaller counterparts and in tilapia it has been noted that fry anaesthetised along with much larger fish recover surprisingly rapidly from the effects of the drug (Ross and Geddes, 1979). The differences noted are probably related to changing liver enzyme activities and fat deposition.

Many drugs such as MS222 and benzocaine may be fat soluble. Thus, in larger, older fish or in gravid females duration of anaesthesia may be prolonged and consequent recovery slower as the drug is slowly removed from the lipid reserves for clearance via the gills or kidney or for metabolic degradation.

In common with all vertebrates, diseased or exhausted animals are very susceptible to anaesthetic treatment. Schoettger and Steuke (1970) showed that pike, *Esox lucius,* and walleye, *Stizostedion vitreum,* depleted after spawning were more susceptible to MS222. Richards (personal communication) found that sea trout infected with *Saprolegnia* did not recover reliably from MS222 or quinaldine anaesthesia.

Fish culturists are often concerned about the repeated use of anaesthetics and their possible effects on fish growth and performance. McFarland and Klontz (1969) claim that the careful and proper use of tertiary amyl alcohol, methyl pentynol and MS222 is unattended by side effects even with repeated use. Nakatani (1962) induced deep MS222 anaesthesia (100 mg.l^{-1}) in trout 5 times per week for 21 weeks and noted no side effects. In addition, Ross and Geddes (1979) described regular use of benzocaine on tilapia with no apparent reduction in growth or spawning ability.

Environmental or abiotic factors affecting anaesthesia

Aquatic invertebrates and fish, with the exception of tunas, are all ectotherms whose body temperatures closely follow that of their environment because of highly efficient heat exchange in the gills and to a lesser extent the skin. Consequently, the effects of temperature on anaesthetic dose can be considerable. Unfortunately, there is no simple underlying relationship and the effect is entirely dependent on the type of drug used. With MS222 and benzocaine, higher doses are required at higher temperatures to produce the same effect. Sehdev *et al.* (1963) showed a similar effect for 2-phenoxyethanol and they also showed that its therapeutic index was increased at lower temperatures. Obviously physico-chemical passage of the drug into the fish is also temperature related.

The pH of an anaesthetic solution will influence its efficacy, possibly by affecting the ratio of charged to uncharged molecules. In addition, a medium of low pH (e.g. unbuffered MS222; pH 3.8 at 30 mg.l^{-1}) induces a stress reaction. Quinaldine, a base with a pKa of 5.42, forms increasingly ionised solutions as the pH falls and loses its anaesthetic efficacy.

Because of the buffering capacity of sea water and its ionic constituents the effects of some drugs may be modified, even in the same species. In general, most anaesthetic drugs are effective in sea water, but the barbiturates are antagonised by high calcium levels (McFarland and Klontz, 1969).

In summary

Although it is useful for operators to be aware of these factors, it is not always possible to take account of all of them in a predictable way. The cumulative effects of these variations are often seen when anaesthetising batches of fish, when it frequently becomes obvious that some individuals are responding in a very different manner to the bulk of their fellows.

References

Houston, A.H., Woods, R.J. (1972). Blood concentrations of tricaine methanesulphonate in Brook Trout, *Salvelinus fontinalis*, during anaesthetization, branchial irrigation and recovery. *Journal of the Fisheries Research Board of Canada.* 29 (9): 1344-1346.

Huish, M.T. (1972). Some responses of the Brown Bullhead (*Ictalurus nebulosus*) to MS222. *Progressive Fish Culturist.* 34 (1): 27-32.

McFarland, W.N. and Klontz, G.W. (1969). Anaesthesia in fishes. *Federal Proceedings.* 28 (4): 1535-1540.

Nakatani, R.E. (1962). A method for force-feeding radio-isotope to yearling trout. *Progressive Fish Culturist.* 24: 156-160.

Ross, L.G. and Geddes, J.A. (1979). Sedation of warm water fish species in aquaculture research. *Aquaculture.* 16: 183-186.

Schoettger, R.A. and Steucke, E.W. (1970). Quinaldine and MS222 as spawning aids for Northern Pike, Muskellunge and Walleyes. *Progressive Fish Culturist.* 82 (4): 199-205.

~~~~

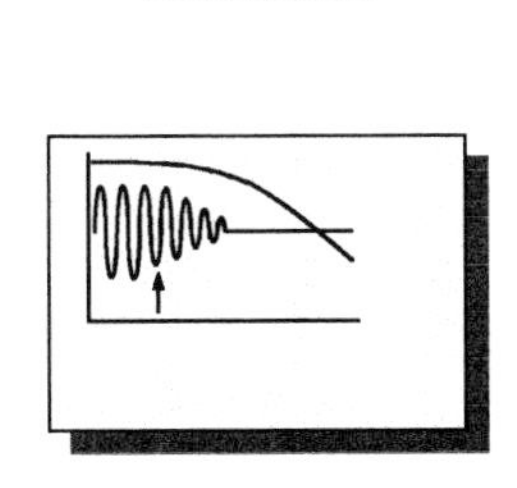
~~~~

Chapter 6
Anaesthesia of Aquatic Invertebrates

Introduction

Invertebrate anaesthesia is not as well developed as fish anaesthesia, one of the principal reasons for this being that the occasions to use it are fewer. Despite this, a wide range of substances has been used from time to time with both molluscs and crustaceans. Clearly, in most aspects of mollusc culture there will be little to no requirement to use anaesthesia. The actions of anaesthetic drugs in molluscs have, however, been extensively studied as these animals form excellent model systems for studying the general mechanisms of drug action. Some isolated preparations from crustaceans have also been used as model systems. Most operations in crustacean culture can also be conducted without anaesthesia, although the motor capabilities of shrimp can present problems in handling. There has, consequently, been some interest in investigating crustacean anaesthetics, particularly for transportation.

Anaesthesia of molluscs

A wide range of simple materials has been used to anaesthetise molluscs, principally for museum work and humane dissection, frequently with no recovery. Smaldon and Lee (1979) note that many of the materials used for such purposes are also effective, reversible anaesthetics, if used at an appropriate dose level. Some guidelines for anaesthetic doses of common materials effective for invertebrates are given by the NRC (1989).

General techniques

Ethanol as a 5 to 10% solution works well with molluscs. Different species respond differently and the agent should be added slowly to the chamber containing the animals. Chlorbutanol (chloretone) is effective at 0.05% in cephalopods, but may be used at up to about 0.5% in many other species. A 0.2% solution of chloral hydrate in sea water is also useful. Chloroform, urethane (1%), 2-phenoxyethanol, the barbiturate nembutal (80 to 350 $mg.l^{-1}$) and the anaesthetics stovaine (amylocaine hydrochloride up to 1%) and eucaine (benzamine hydrochloride) also seem to be effective in molluscs (Smaldon and Lee, 1979).

Clove oil is effective as a muscle relaxant and anaesthetic in a range of molluscs (Araujo *et al.* 1995).

Carbon dioxide, most frequently dispensed in the form of soda water, can be used effectively, adding it to the animal chamber as required.

Magnesium chloride or magnesium sulphate are probably the oldest immobilising agents used with invertebrate animals. Magnesium ions are effective because they block muscle action, competing with the calcium required for synaptic transmission. They can be used as a saturated solution in distilled water for freshwater animals and can be diluted with an equal volume of sea water for use with marine animals. A more general procedure is to add iso-osmotic 7.5% w/v magnesium chloride, $MgCl_2.6H_2O$, to the medium containing the animal, monitoring the dose level by noting when the animal does not respond to prodding. It should be noted that these agents immobilise and probably do not provide real anaesthesia and their analgesic properties are unknown.

As may be expected in ectotherms, careful use of cooling produces effective immobilisation, although the true degree of analgesia and anaesthesia is suspect. Many molluscs may be safely cooled to about 2°C, or even to 0°C in some species. Long periods at low temperatures may cause death, especially in warm water species.

Throughout any procedure, some attention must be given to managing the anaesthesia. Water quality should be maintained by oxygenation and filtration. A good oxygen supply with metabolite removal will be particularly important in the case of cephalopods. The temperature should be controlled, or at least managed so that animals do not overheat. Steps may also be needed to maintain a damp skin or exoskeleton, usually by spraying with water or sea water.

Bivalves

Only a limited amount of work has been done specifically on bivalve anaesthesia. 2-Phenoxyethanol at up to a 1% solution is effective in clams (Owen 1955). Culloty and Mulcahy (1992) reviewed the effectiveness of nembutal, magnesium chloride, magnesium sulphate, chloral hydrate, urethane, menthol and benzocaine in the flat oyster, *Ostrea edulis,* at a range of concentrations. They judged anaesthesia to have occurred when the valves open but do not close when the animal is touched, the usual bivalve response. Nembutal (0.012 to 0.12%), menthol (up to 2%) and urethane (0.3 to 2.0%) gave very poor anaesthesia, and magnesium sulphate required a 30% solution to be effective. Two to five per cent chloral hydrate was effective, as was a 0.1% solution of benzocaine. Overall, they concluded that 3.5% w/v magnesium chloride was the agent of choice for oysters, inducing anaesthesia relatively quickly, lasting for up to 90 minutes and allowing rapid recovery with minimal stress and mortality. Heasman *et al.* (1995) also obtained useful anaesthesia of *Pecten fumatus* using 30 $g.l^{-1}$ magnesium chloride, but considered that magnesium sulphate induced excessive mortalities. The same authors successfully induced anaesthesia in *Pecten fumatus* using chloral hydrate at 4 $g.l^{-1}$. Ehtesami (1993) reports effective anaesthesia of *Pinctada radiata* using MS222 at 1 $mg.l^{-1}$.

Gastropods

Lymnaea stagnalis has been used as a model to study general anaesthesia as the snail is reversibly anaesthetised by the fluoranes (Girdlestone *et al.*, 1989). Isolated preparations of its nervous system have also provided good model systems for studying mechanisms of anaesthetic action (Girdlestone, McCrohan and Winlow, 1989). It has large, uniquely identifiable nerve cells, some of which have identified neurotransmitters and simple, monosynaptic connections to other cells, making an ideal study system (Winlow *et al.*, 1992). Similar studies have been conducted using neurones from *Aplysia* and *Helix* (Parmentier *et al.*, 1979). *Lymnaea stagnalis* can be reversibly anaesthetised by the gases halothane, enflurane and isoflurane, the effective doses being 0.83%, 1.01%, and 1.09% v/v, respectively (Girdlestone *et al.*, 1989).

In addition to immersion as described earlier, magnesium chloride can be injected as a 10% solution close to the cerebral ganglia, giving quick relaxation for 5 to 15 minutes (Runham *et al.*, 1965). Gastropods can also be reversibly anaesthetised by immersion in solutions of 0.1% nembutal or 0.3% MS222.

Injection of the acetylcholine analogue, succinyl choline, dissolved in sea water at 5 $g.l^{-1}$ into the space near the cerebral ganglia at 0.5 mg per 10 g live weight has been shown to give good relaxation in many gastropods, including species of *Aplysia*,

Aeolidia, Hermissenda, Strombus, Bulla, Pleurobranchia and *Acmaea* (Beeman, 1969). For research work, injection appears to give more controllable results than immersion.

Cooling in a refrigerator can provide effective immobilisation. However, true anaesthesia is probably not achieved and little analgesia may be provided. *Littorina rudis* chilled to below zero can recover after several days with a high percentage survival (B. Ross, unpublished data).

White *et al.* (1996), working with *Haliotis midae* of up to 90 mm shell length, obtained useful anaesthesia using magnesium sulphate (4 to 22 g.100.cm^{-3}) and 2-phenoxyethanol (0.05 to 0.3 cm^{3}.100 cm^{-3}).

Cephalopods

The cephalopods have not yet been cultured although there is considerable commercial interest to do so. They are more widely used as experimental animals, and because of their extremely advanced nervous system and brain, they have in recent years been accorded equal status with vertebrate animals by requiring that such experimental work is properly licensed. They are particularly difficult to handle as they are not only agile and fast, but also have a very delicate integument which is loosely attached to the underlying body. Gleadall (1991) used a standard procedure to screen the potential of 11 anaesthetic agents for use with *Octopus* sp., having noted that ethanol and urethane are ineffective at temperatures below 12°C. He compared ethanol, urethane, magnesium chloride, MS222, metomidate, propoxate, chloral hydrate, chlorbutanol, menthol, nicotine sulphate and 2-phenoxyethanol, as well as cooling techniques. He showed that some materials were ineffective, toxic or fatal and concluded that properly controlled anaesthesia was not really achieved with any of these agents.

Induction of anaesthesia in *Octopus* sp. using 7.5% magnesium chloride in distilled water, mixed with an equal volume of sea water for use, requires about 13 minutes and recovery about 8 minutes (Best and Wells, 1983). Octopuses cooled to 3 to 5°C from 24°C are immobilised well, but with less effective muscle relaxation (Andrews and Tansey, 1981).

In squid, urethane is effective at 3% in sea water, providing handling ability after only a few minutes exposure and a recovery period of 3 to 15 minutes (O'Dor *et al.,* 1977), although urethane is now considered an unsuitable material because of its carcinogenic properties. Two per cent ethanol in sea water is also effective. It should be noted that both of these materials cause initial hyperactivity, which can be traumatic.

O'Dor *et al.* (1990) recommend the use of 7.5% magnesium chloride in distilled water for squid, mixed with an equal volume of sea water for use. Garcia-Franco (1992) found that only 1.5 to 2% magnesium chloride, 3 to 4% magnesium sulphate or 1 to 3% ethanol produced effective anaesthesia in the squid, *Sepioteuthis sepioidea*. Many other agents are ineffective, possibly related to the nature of the nerve transmitters and receptors in these molluscs. O'Dor *et al.* (1990) anaesthetised the squid *Illex illecebrosus* and *Loligo pealei* for brief procedures such as weighing or tattooing using 1.5% ethanol, but cautioned that mortality is high if animals are left in the solution for long periods. Ethanol or magnesium anaesthesia combined with cooling to 5 to 7°C is better for surgery lasting up to 15 minutes. Assisted ventilation is obviously necessary during such longer procedures with cephalopods and cooled, oxygenated anaesthetic should be passed through the mantle cavity continuously.

Anaesthesia of crustaceans

In addition to the standard acetylcholine and noradrenaline transmitters used in nerve cells, crustaceans use 5-hydroxytryptamine, glutamate and possibly polypeptide-like transmitter substances. It is perhaps not surprising that crustaceans respond differently to certain anaesthetic agents, possibly because their synaptic receptor sites are not affected by some of the commonly used drugs.

General techniques

As in molluscs, a wide range of anaesthetic agents has been used. Ethanol as a 5 to 10% solution works well. Different species respond differently and the agent should be added slowly to the chamber containing the animals. Chlorbutanol is effective at as little as 0.05% with some crustaceans, but may be used at up to about 0.5% in other species. Chloroform (1%), urethane (1%), 2-phenoxyethanol (1.5%), ethane disulphonate (2.5 $g.l^{-1}$), isobutyl alcohol (1.5 to 7 $cm^3.l^{-1}$), methyl pentynol (5 $cm^3.l^{-1}$) and MS222 (0.5 $g.l^{-1}$) can also be effective (Smaldon and Lee, 1979).

As in molluscs and many other animals, carbon dioxide is an effective anaesthetic. It is most frequently dispensed in the form of soda water, and is added to the chamber containing the animals in sea water in approximately equal parts until the desired effect is obtained.

As described for molluscs, magnesium chloride or magnesium sulphate are effective immobilising agents, although there is probably little accompanying analgesia. A saturated solution in distilled water has been used for freshwater animals and this can be diluted with an equal volume of sea-water for use with marine animals. Where recovery is required, a more general procedure is to add isosmotic 7.5% w/v magnesium chloride, $MgCl_2.6H_2O$, to the medium containing the animal, monitoring the dose level by noting when the animal does not respond to prodding.

Cooling is also an effective means of immobilisation for crustaceans, although the true degree of analgesia and anaesthesia is unknown. Many crustaceans may be safely cooled to 1 to 2°C, although it should be noted that autotomy of appendages may occur in some species.

Throughout any procedure, some attention must be given to managing the anaesthesia. Water quality should be maintained by oxygenation and filtration. This will be particularly so in the case of the more active species. The temperature should be controlled, or at least managed so that animals do not overheat. Steps may also be needed to maintain a damp exoskeleton, usually by spraying with water or sea water.

Branchiopods

Daphnia magna was shown to be anaesthetised by methoxyflurane, halothane, isoflurane and enflurane when concentrations of 0.095, 1.005, 1.170 and 1.42% w/v in 100% oxygen, respectively, were bubbled through the water (McKenzie, Calow and Nimmo, 1992). The effect was temperature dependent (McKenzie *et al.*, 1992).

Stomatopods

Ferrero and Pressacco (1982) administered isobutanol to the stomatopod *Squilla mantis* by cardiac injection and showed that it was reliable and effective at 200 $\mu l.kg^{-1}$ at temperatures between 11 and 17°C. Xylazine, administered by the same route was effective at 100 $mg.kg^{-1}$, but caused some mortalities. It is thus not considered to be a realistic option for these crustaceans.

Amphipods

Freshwater shrimp, *Gammarus pulex*, (Simon *et al.*, 1983; Smith *et al.*, 1984) have been used to study the ameliorating effect of pressure upon anaesthesia, again in an attempt to shed light on the site of action of these agents. Ahmad (1969) used MS222 at 0.5 to 1.0 $g.l^{-1}$ to successfully anaesthetise *G. pulex* and noted that induction was slower but that the safety margin was greater at lower temperatures. Species of *Corophium* can also be anaesthetised by MS222 at this dose level (Gamble, 1969).

Decapods

Isolated crayfish and crab axons have been used as models in elucidating the general mechanisms of anaesthetic action, particularly with general anaesthetic agents used in higher animals. Such isolated preparations are effective in identifying whether anaesthetics act by a general action on the lipid bilayers of nerve cells or whether more specific target receptors are involved.

Many decapods have a reflex action of grasping an object, for example a wet cloth or tissue towel, which is presented to them. They will frequently hold this object tenaciously for some time and, assuming that no anaesthesia or analgesia is necessary, this may be sufficient to carry out many simple procedures. *Carcinus maenas* "immobilised" in this way were sufficiently distracted to enable the carapace to be drilled to fit near-heart electrodes and to have a small telemetry tag connected and cemented onto their backs (Fig. 15). The whole procedure lasted about 10 minutes and no drugs were used at any stage.

In general, decapods may be anaesthetised by exposure to a gaseous anaesthetic agent dispersed in air, by immersion in a solution or by injection.

Obradovic (1986) used halothane at 0.5% by volume in air to effectively anaesthetise the crayfish *Astacus astacus*. This technique would certainly work with other decapods, although it has not been widely used.

Foley *et al.* (1966) found that immersion in isobutyl alcohol at concentrations of 0.5 to 14.4 $cm^3.l^{-1}$ anaesthetised the lobster *Homarus americanus*. After brief hyperactivity, ataxia was produced and recovery from this state took 10 minutes in clean sea water. MS222 gives varied results. *Crangon septemspinosa* is anaesthetised by 0.5 g.l^{-1} MS222. By contrast, Obradovic (1986) showed that MS222 immersion was ineffective in the crayfish *Astacus astacus* at low concentrations (100 mg.l^{-1}) and had only a mild effect at high concentration (1000 mg.l^{-1}) in these animals. The latter concentration would be fatal in fish and this perhaps reflects the difference in neurotransmitters found in crustaceans.

Fig. 15. Ultrasonic heart rate telemetry tag fitted to the carapace *of Carcinus maenas* during a procedure using no anaesthetic. The crab lived for many days before leaving the range of the receiving equipment (L.G. Ross, unpublished data).

Injectable anaesthesia is possible in Decapoda and the most comprehensive study in the crabs *Cancer* and *Carcinus* was by Oswald (1977). He tested a range of materials by injection into the haemocoel via the arthrodial membrane of a posterior leg. There are, of course, many similarly structured, alternative routes to the haemocoel, depending upon species (Fig. 16). Oswald showed that MS222, benzocaine, lignocaine, d-tubocurarine, gallamine, suxamethonium, decamethonium and guaiacol glyceryl ester were all ineffective and that chlorpromazine induced massive autotomy of limbs. Propanidid (100 mg.kg^{-1}), xylazine (70 mg.kg^{-1}), and pentobarbitone (250 mg.kg^{-1}), were all useful injectable drugs, giving sleep times of 60, 45 and 90 minutes respectively. He concluded that alphaxolone-alphadolone (Saffan) at 30 mg.kg^{-1} and procaine at 25 mg.kg^{-1} were most effective. Procaine produced a brief initial hyperactivity followed by anaesthesia within 30 seconds of injection into crabs and sleep was maintained for 2 to 3 hours.

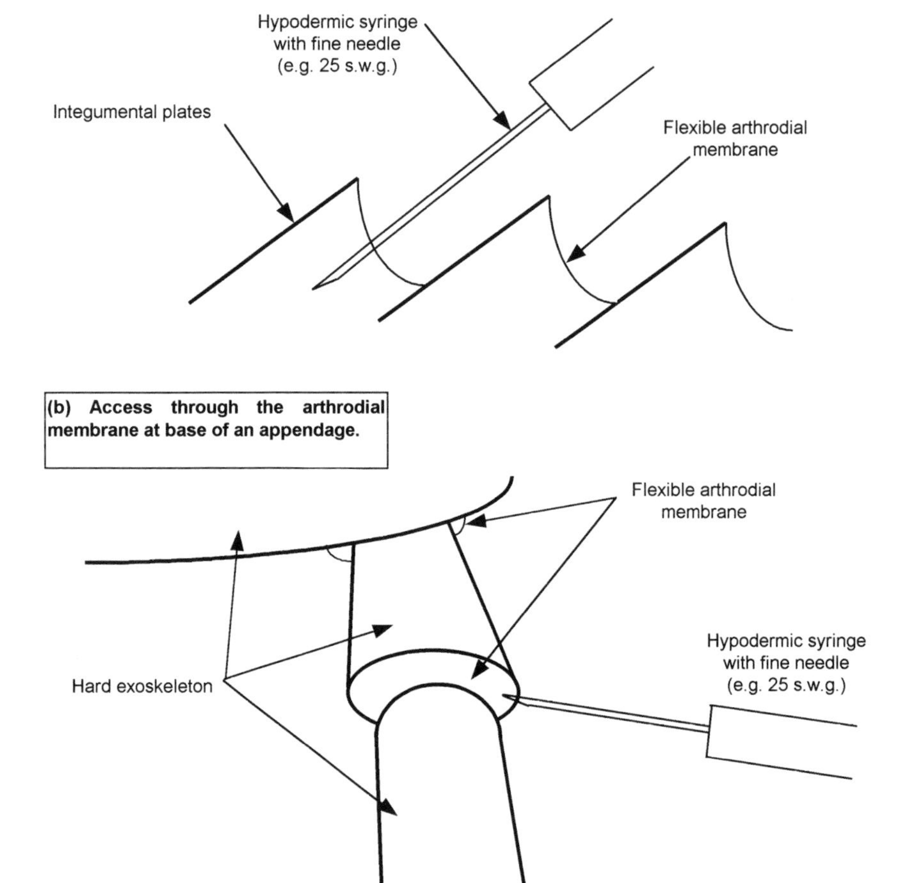

Fig. 16. Some sites for the injection of drugs into crustaceans through the arthrodial membrane.

In summary

In aquaculture, the need to anaesthetise molluscs or crustaceans will be quite rare, most procedures in husbandry and management being possible without additional aids. However, the ethics of procedures such as the unanaesthetised ablation or cautery of crustacean eyestalks may be questionable. Overall, although sedation and anaesthesia of these groups has not been as extensively investigated as in fish, there are a range of techniques and drugs available which will give good results. Caution should be taken in using any technique with a previously untested species of mollusc or crustacean as the variability in response in these groups can be wide and unpredictable.

References

Ahmad. M.F. (1969). Anaesthetic effects of tricaine methane sulphonate (MS222 Sandoz) on *Gammarus pulex* (L.) (amphipoda). *Crustaceana (Leiden).* 17: 197-201

Andrews, P.L.R. and Tansey, E.M. (1981). The effects of some anaesthetic agents in *Octopus vulgaris. Comparative Biochemistry and Physiology.* 70C: 241-247.

Araujo, R., Remon, J.M., Moreno, D. and Ramos, M.A. (1995). Relaxing techniques for freshwater molluscs: Trials for evaluation of different methods. *Malacologia.* 36 (1-2): 29-41.

Beeman. R.D. (1969). The use of succinylcholine and other drugs for anaesthetising or narcotising gastropod molluscs. *Publications of the Zoological Station, Naples.* 36: 267-270.

Best, E.M.H. and Wells, M.J. (1983). The control of digestion in *Octopus.* I: The anticipatory response and effects of severing the nerves to the gut. *Vie Milieu.* 33: 135-142.

Culloty, S.C. and Mulcahy, M.F. (1992). An evaluation of anaesthetics for *Ostrea edulis* (L). *Aquaculture.* 107 (2-3): 249-252.

Ehteshami, F. (1993). Anaesthetising *Pinctada radiata* with MS 222. *Iranian Fisheries Bulletin.* 3: 1.

Ferrero, E.A. and Pressacco, L. (1982). Anaesthetic procedures for crustacea. An assessment of isobutanol and xylazine as general anaesthetics for *Squilla mantis* (Stomatopoda). *Memorie di Biologia marina e di Oceanografia.* 12 (1): 47-75.

Gamble. J.C. (1969). An anaesthetic for *Corophium volutator* (Pallas) and *Marinogammarus obtustatus* (Dahl), Crustacea, Amphipoda. *Experientia.* 25: 539-540.

Garcia-Franco, M. (1992). Anaesthetics for the squid *Sepioteuthis sepiodea* (Mollusca: Cephalopoda). *Comparative Biochemistry and Physiology.* 103C (1): 121-123.

Girdlestone, D., Cruikshank, S.G.H and Winlow, W. (1989). The actions of three volatile general anaesthetics on withdrawal responses of the pond snail *Lymnea stagnalis* (L). *Comparative Biochemistry and Physiology*. 92C (1): 39-43.

Girdlestone, D., McCrohan, C.R. and Winlow, W. (1989). The actions of halothane on spontaneous activity, action potential shape and synaptic connections of the giant serotonin-containing neurone of *Lymnea stagnalis* (L). *Comparative Biochemistry and Physiology*. 93C (2): 333-339.

Gleadall, I.G. (1991). Comparison of anaesthetics for octopuses. *Bulletin of Marine Science*. 49 (1-2): 663.

Heasman, M.P., O'Connor, W.A. and Frazer, A.W.J. (1995). Induction of anaesthesia in the commercial scallop *Pecten fumatus*. *Aquaculture*. 131 (3-4): 231-238.

McKenzie, J.D., Calow, P. and Nimmo, W.S. (1992). Effects of general anaesthetics on intact *Daphnia magna* (Cladocera: Crustacea). *Comparative Biochemistry and Physiology*. 101C (1): 9-13.

McKenzie, J.D., Calow, P., Clyde, J., Miles, A., Dickinson, R., Lieb, W.R. and Franks, N.P. (1992). Effects of temperature on the potency of halothane, enflurane and ethanol in *Daphnia magna* (Cladocera: Crustacea). *Comparative Biochemistry and Physiology*. 101C (1): 15-19.

NRC (1989). Laboratory animal management. *Marine Invertebrates*. National Academy Press, Washington D.C. 382pp.

Obradovic, J. 1986. Effects of anaesthetics (halothane and MS222) on crayfish, *Astacus astacus*. *Aquaculture*. 52 (3): 213-217.

O'Dor, R.K., Durward, R.D. and Balch, N. (1977). maintenance and maturation of squid (*Illex illecerebrosus*) in a 15 metre circular pool. *Biological Bulletins of Woods Hole. Massachusetts*. 153: 322-335.

O'Dor, R.K., Portner, H.O. and Shadwick, R.E. (1990). Squid as elite athletes: Locomotory, respiratory and circulatory integration. In: *Squid as Experimental Animals* (Edited by D.L. Gilbert, W.J. Adelman and J.M. Arnold). Plenum Press, New York. 516pp.

Oswald, R.L. (1977). Immobilisation of decapod crustacea for experimental procedures. *Journal of the Marine Biological Association of the U.K.* 57: 715-721.

Parmentier, J.L., Shrivastav, B.B., Bennett, P.B. and Wilson, K.M. (1979). Effect of interaction of volatile anaesthetics and high hydrostatic pressure on central neurones. *Undersea Biomedical Research*. 6 (1): 75-91.

Simon, S.A., Parmentier, J.L. and Bennett, P.B. (1983). Anaesthetic antagonism of the effects of high hydrostatic pressure on locomotory activity of the brine shrimp *Artemia*. *Comparative Biochemistry and Physiology*. 75A (2): 193-199.
Smaldon. G. and Lee, E.W. (1979). *A Synopsis of Methods for the Narcotisation of Marine Invertebrates*. Royal Scottish Museum Information Series. No. 6. Edinburgh.

Smith, E.B., Bowser-Riley, F., Daniels, S., Dunbar, I.T., Harrison, C.B. and Paton, W.D.M. (1984). Species variation and the mechanism of pressure-anaesthetic interactions. *Nature*. 311 (5981): 56-57.

White, H.I., Hecht, T. and Potgeiter, B. (1996). The effect of four anaesthetics on *Haliotis midae* and their suitability for application in commercial abalone culture. *Aquaculture*. 140 (1-2): 145-151.

Winlow, W., Yar, T., Spencer, G, Girdlestone, D. and Hancox, J. (1992). Differential effects of general anaesthetics on identified molluscan neurones *in situ* and in culture. *General Pharmacology*. 23 (6): 985-992.

~~~~

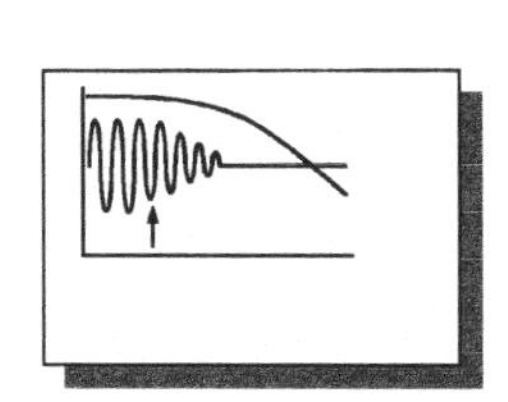
~~~~

Chapter 7
Anaesthesia of Fish
I. Inhalation Anaesthesia

Introduction

This most widely used technique depends on the anaesthetic drug being in aqueous solution. It is ventilated (inhaled) by the fish and rapidly enters the arterial blood from where it is a very short route to the central nervous system (Fig. 17). This is analogous to gaseous anaesthesia in terrestrial vertebrates. On return to fresh water the drugs, or their metabolites, are most frequently excreted via the gills and to a much lesser extent via the skin. Some materials may be excreted via the kidney, either intact or as metabolites.

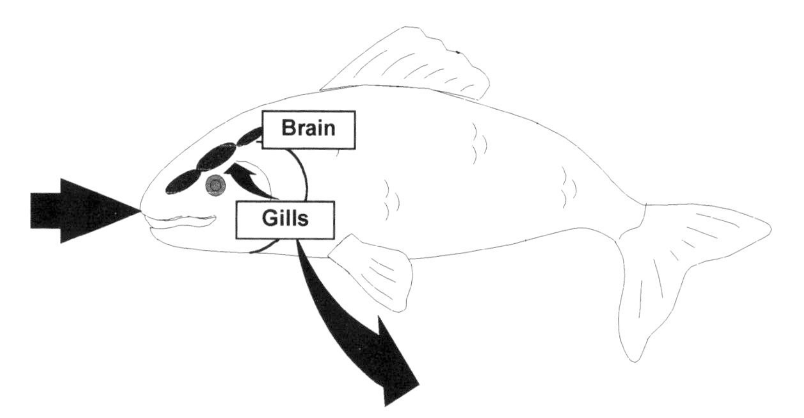

Fig. 17. Route of inhaled drugs to the central nervous system of fish.

There are certain species of fish which can breathe air, rendering this technique unsatisfactory. The snakehead, *Ophiocephalus* sp., is an example of an obligate air breather and its gills are much reduced, being used only for excretion and osmoregulatory purposes. Consequently, although the animal does respond to inhalation anaesthesia, induction is usually lengthy, unpredictable and therefore frustrating. Similar problems are encountered with the clariid catfish for the same reasons. The European eel, *Anguilla anguilla*, and most catfish are facultative air breathers and when placed in noxious solutions have a tendency, and the ability, to hold their breath. In this case, although induction will be lengthy, it is nevertheless effective.

Water quality maintenance during inhalation anaesthesia

It should be obvious that water quality needs to be carefully controlled during inhalation anaesthesia, particularly where large numbers of animals are being handled and where baths are being reused. This is part of the general management of the anaesthetic procedure and the level of anaesthesia where it is maintained for more than a brief period. The main problems are those facing all aquatic animals: temperature, dissolved oxygen concentration, ammonia levels and build up of faecal and other solids in the baths.

The stock baths and recovery vessels should all use water from which the animals originate and at the temperature to which the animals are acclimated. Small electrical heaters and coolers can be used if needed, although this will rarely be the case in short procedures. Cooling can also be effected with ice. Care should be taken that animals do not overheat or overcool while out of the water for handling. In the extreme environmental conditions of the tropics and subtropics or northern winters this can occur very quickly indeed.

The stock containers, anaesthetic bath and recovery vessel should be equipped with airstones or diffusers. In all cases, aeration is best if it can be effected using a diffuser, as the gas exchange efficiency from the finer bubbles produced is much greater than if a simple airstone is used. It is also prudent to ensure that high-pressure compressors do not issue vaporised oil into the water and for this reason it is always best to use low-pressure air blowers if available.

Excreted solids are more difficult to deal with, but water changes can be used and should be effected as often as needed by visual inspection. The anaesthetic bath is more costly to replace and if large numbers of animals are to be used some simple filtration system should be used. In general, however, working solution replacement is frequently the only practical means.

Ammonia levels can be reduced to some extent by continuously passing the water over zeolites, clinoptilolites or other ion-exchange materials. This extra level of complexity is rarely necessary, however, and the lower metabolic rate induced by the anaesthesia itself will be of some assistance.

The basic procedure

For simple procedures it is usually possible to immerse the fish directly in a suitable concentration of drug so that spontaneous ventilation is maintained, and this approach is widely use in fish farming. The simplest method of achieving this is to make up the required drug concentration in an aerated container and to quickly but gently transfer the fish to the container. For most procedures in aquaculture and fisheries management this technique is quite adequate. Induction should be rapid, handling time will be minimal and the fish can be transferred to well-aerated clean water within a few minutes, having been weighed, measured, marked or whatever is required (Fig. 18). Using this procedure it is usually not necessary to exceed stage 1 and plane 2 of McFarland's scheme (Table 5) for adequate handling. Consequently, ventilation of the gills is spontaneously maintained and use of a well-aerated drug solution and recovery baths will ensure an adequate oxygen supply and mortality will be the exception.

Using clean, inexpensive containers, this technique and scheme can easily be adapted for use in the laboratory, on the farm or in the field. Figure 19 shows a rainbow trout being fitted with ultrasonic tracking tags in the field, following simple benzocaine anaesthesia.

For successful bulk use of this approach, a scheme of work is useful to ensure that all concerned adhere to a procedure that will safeguard the welfare of the animals. The box on page 63 gives an example of a specific scheme for inhalation anaesthesia using benzocaine and outlines the procedure to be followed for simple handling and where no artificial ventilation is required.

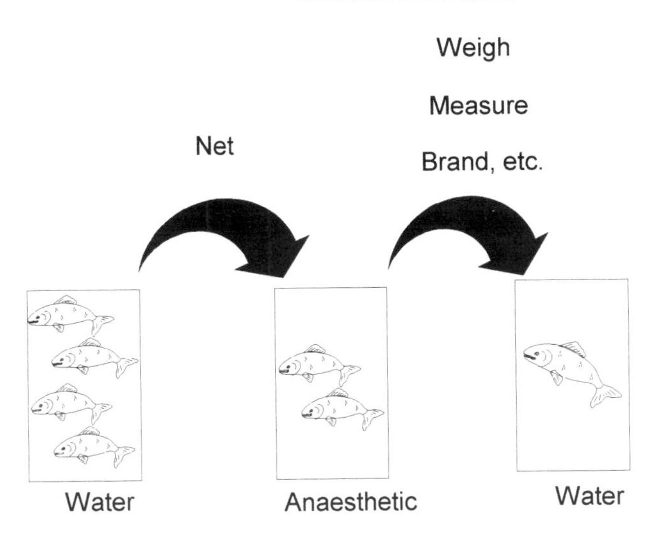

Fig. 18. Schematic representation of simple inhalation anaesthesia procedure as used in fisheries and aquaculture.

Fig. 19. Rainbow trout being fitted with ultrasonic tracking tags in the field, following simple benzocaine anaesthesia.

Bulk Inhalation Anaesthesia: Scheme of Work

1. Prepare a stock solution of benzocaine, usually 100 g benzocaine dissolved in 1 litre of ethanol or acetone. Store in a dark bottle and refrigerate, if possible.

2. Prepare the anaesthetic bath to the appropriate final concentration (see Table 10, page 75) by slowly adding stock solution to water with **thorough stirring** to prevent the drug coming out of solution. Stirring is very important at this stage.

3. Prepare a recovery vessel of clean, well-oxygenated water.

4. Ensure that the stock tank, anaesthetic bath and recovery vessel are all at the same temperature. Adjust if necessary.

5. Ensure that all relevant equipment for the intended technique, nets, buckets, syringes, needles, injectable materials, tags, tissue wipes, etc., is prepared and to hand before proceeding any further.

6. Quickly, but gently, net out a small batch of fish from stock and transfer them immediately to the anaesthetic bath. The number of fish per batch will depend on the time taken to handle each fish. Even experienced workers should limit themselves to a very small batch in the first netting, say 3 to 4 animals, to allow for set-up errors and to regain handling confidence. On no account should the fish remain in the bath for extended periods. As a rough guide fish should not remain in any anaesthetic bath for longer than 10 minutes.

7. For simple procedures it is usually possible to achieve adequate handling even before the fish have lost equilibrium. Operations should commence as soon as a fish can be picked up, gently and without struggling.

8. After handling, the fish should be placed immediately in the recovery tank. Substantial recovery will normally occur within 1 minute, although some species may require slightly longer. On no account should the recovery tank be allowed to become over-stocked.

Unfortunately it is difficult to maintain a uniform depth of anaesthesia using this technique. Remember that drug dose and exposure time may act cumulatively to determine the final stage of anaesthesia reached. For example Houston *et al.* (1971) have shown that levels of MS222 in brain and muscle continue to increase after blood levels have attained equilibrium. Consequently, a drug dose which is initially satisfactory can produce progressively deeper anaesthesia and eventual ventilatory arrest. The resulting decline of water flow in the buccal cavity contributes to a reflex decline in heart rate and dorsal aortic blood pressure. A progressive hypoxia ensues which is further complicated in some cases by swelling of erythrocytes causing a decrease in gill capillary blood flows (Soivio *et al.*, 1977). Species with a high body lipid content, or large mature fish, retain fat-soluble anaesthetic agents for a longer period after recovery.

During simple procedures these complications will not occur, but it will readily be appreciated that lengthy exposure to a dissolved anaesthetic agent should be avoided if progress of the animals through the successive stages of anaesthesia is to be prevented. Where an animal fails to recover spontaneous ventilation, the water flow-heart rate reflex can be exploited. As noted earlier, a reduction in buccal water flow causes a reflex reduction in heart rate. Fortunately, in the anaesthetised animal the reverse is also true, in that increasing water flow through the buccal cavity will accelerate and regularise heart rate markedly. This increases gill blood flow and eliminates the drug more rapidly, hastening recovery. In emergencies this can be achieved by moving the fish backwards and forwards in the recovery bath, or by **gently** passing water through the buccal cavity with a narrow hose. Once the recovery is sufficient for the fish to ventilate spontaneously, it can be left to complete the process unattended. During this phase, ventilation can often be very deep with powerful, regular movements of the opercula.

Direct application to the gills

Some workers have used sprays or atomiser bottles to apply drugs directly to the gills in larger fish. This can be useful with large animals where immersion is impractical. The earliest accounts of this were by Gilbert and Wood (1957), who anaesthetised large sharks by holding back the head with a gaff or rope and spraying the gills directly with 1000 $mg.l^{-1}$ MS222. They suggested that a water pistol, syringe pump or spray bottle could be used but small (say 1 litre) plastic spray bottles are now very inexpensive and widely available. The sharks were maintained in anaesthesia by further spray applications of MS222, or the procedure could be shortened by spraying with drug-free water. Kaneko (1982) describes problems in handling large specimens of air-breathing *Arapaima* and *Lepisosteus*, which were incompletely anaesthetised by immersion in 10% quinaldine. He successfully used direct sprays to apply 200 $mg.l^{-1}$ MS222 or 20% Fluothane on to the gills. Kidd and Banks (1990) described the use of this method for stripping broodstock cutthroat trout, *Salvelinus namaycush*. They note that the initial handling was possible without the need for drug immersion and that there was no effect on subsequent egg hatching success. Indeed, this approach may be a useful way to isolate eggs from any possible harmful effects of anaesthetic drugs.

Artificially ventilated inhalation anaesthesia

In more complex work, where ventilatory arrest is unavoidable, a system of artificial ventilation should always be used. Numerous descriptions of appropriate systems for ventilated anaesthesia exist in the literature (Bell, 1964; Brown, 1987). All that is required is a supply of aerated anaesthetic solution at the correct temperature delivered from a pipe or mouthpiece into the buccal cavity, a collection system for the effluent anaesthetic and preferably a recycling system for used solution. The anaesthetic should be aerated to ensure near-saturation of oxygen and to remove dissolved CO_2. It should also be maintained at the correct temperature using cooling or heating, as required (Figs 20 and 21).

The fish is first sedated in a container of appropriate anaesthetic solution, as described in the basic procedure, above. The fish is then transferred to a suitable fish holder on an apparatus containing the maintenance solution, which will usually be at a lower concentration than that used for induction. The anaesthetic solution is introduced into the buccal cavity from a plastic or rubber tube. If available, silicon rubber is best as it is softer. A useful fish holder can be made by slitting a block of foam plastic half-way through (Fig. 22). When this block is opened, it will grip the animal lightly while simultaneously allowing water or anaesthetic solution to flow across the gills. It is most convenient to cover the fish holder with some soft, wettable, disposable material which does not readily disintegrate. Similarly wetted operation cloths may be desirable for shielding the fish from the heat of lights, soldering equipment, etc. For management of anaesthesia it is also helpful to use a routine ECG monitor, for example a physiological pre-amplifier and oscilloscope or else a much simpler bleeper triggered by the QRS complex (Ross and Wiewiorka, 1977), to track progress of the subject. It is easy to detect a simple heart rate signal by using clip electrodes attached to the fins.

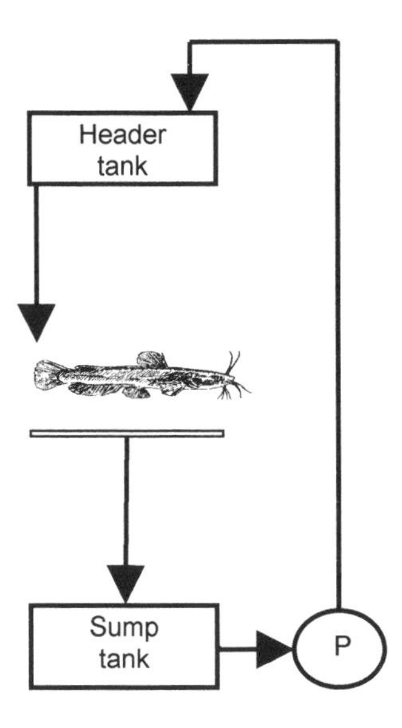

Fig 20. Schematic diagram of simple ventilated anaesthesia system.
P = pump.

Depending on the drugs used and assuming good environmental control, fish can be held in this state for several hours. Clearly, in this situation the two prime complicating factors are body temperature maintenance and prevention of excessive drying of the skin. The former is usually alleviated by adequate control of the water temperature, and the latter by spraying the skin surface with water at regular intervals from a wash bottle or, better still, a small portable atomiser.

Fig. 21. Ventilated anaesthesia system used in the authors' laboratory. Note the cooler unit and header tank at the left, the anaesthetic delivery tube and the foam block fish holder.

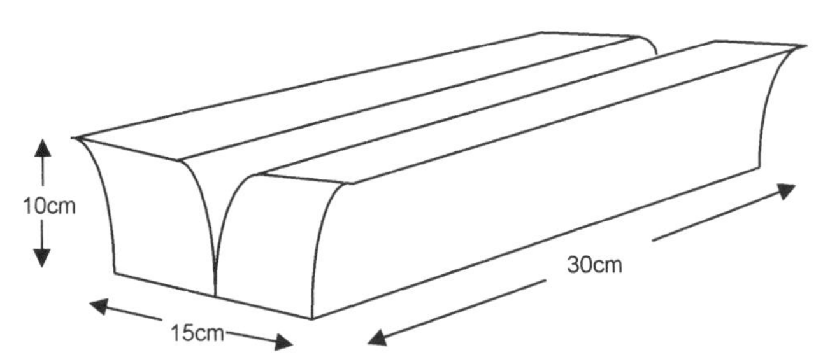

Fig. 22. Simple fish holder made from a foam plastic block, cut part-way through. Sizes will vary according to fish size. In use the holder is covered with sheets of disposable tissue.

To terminate anaesthesia the drug supply is stopped and clean, drug-free water is passed over the gills until spontaneous ventilation returns. At the first appearance of locomotor activity, the fish may be returned to the stock tank. Those wishing to anaesthetise fish of any value should always anticipate ventilatory and/or cardiac arrest and their ensuing complications. Again, it is helpful to remember the ventilatory/cardiac reflex in this context and by simply irrigating the buccal cavity continuously with water, heart rate can be restored and maintained and excretion of drugs and metabolites back across the gills can be considerably speeded up.

Widely used drugs for inhalation anaesthesia

An extensive range of drugs has been used in fish anaesthesia. The more detailed descriptions which follow are limited to those substances currently in the widest use. Dose rates of these drugs for a range of different fish species are summarised in Table 9, the examples having been chosen to illustrate the range of doses useable or the variation which may be encountered.

MS222 (ethyl-m-aminobenzoate, tricaine methane sulphonate)

This drug was first developed by Sandoz in a search for cocaine substitutes. It was initially used as a local anaesthetic, but its effectiveness with ectotherms was recognised and it has since been well investigated with many species. It is a white, crystalline powder which keeps well when dry. It has a sulphonated side-chain and this makes it acidic, but as a consequence it is 250 times more soluble in water than benzocaine. It can be dissolved in fresh water and sea water up to 11% w/v. The low pH solution produced is irritant to fish, however, and a formidable list of physiological consequences of its use have been documented, including elevated haematocrit, erythrocyte swelling, hypoxia, hypercapnia, hyperglycaemia, changes in blood electrolytes, hormone levels, cholesterol, urea, lactate and interrenal ascorbic acid. It should be borne in mind that handling alone can cause some of these changes.

Stock solutions can be prepared, usually 10 $g.l^{-1}$, and these are stable if kept in a dark, stoppered container. Solutions can remain effective for up to 3 months if kept cool and in the dark, but eventually they will darken and slowly lose their potency. The stock solutions or the working solution can be buffered using sodium bicarbonate or Tris-buffer, usually to pH 7.0 to 7.5. Smit *et al.* (1977) recommend that distilled or deionized water should not be used, as neither possesses any buffering capacity.
Induction is rapid and can take as little as 15 seconds. In salmonids it is quickly effective by immersion at about 50 $mg.l^{-1}$, although maintenance levels can then be as low as 10 $mg.l^{-1}$ (Laird and Oswald, 1975). Up to 100 $mg.l^{-1}$ may be required for

tilapias and *Clarias* sp. (Ross and Geddes, 1979). Recovery times are usually rapid and equilibrium and motor activity can be expected to return after only a few minutes.

MS222 has a good safety margin to fish. In trout, Bové (1962) noted that the 30 minute LC_{50} is 82 $mg.l^{-1}$ and the maximum concentration tolerated is a little lower at 63 $mg.l^{-1}$. The effective concentration (at which anaesthesia is produced in 99% of fish in 3 to 4 minutes) is 40 $mg.l^{-1}$ and so the therapeutic index is then 63/40=1.57. This narrows, however, as temperature rises and appears to be less in smaller fish. The drug is more potent in warm waters with low hardness.

MS222 is not currently known to be toxic to humans at the concentrations used.

MS222 is excreted via the urine within 24 hours and tissue levels decline to almost zero in the same time. The withdrawal time required by the FDA is 21 days.

Benzocaine (ethyl-p-aminobenzoate)

Benzocaine is a white crystalline material chemically very similar to MS222. As it does not have the sulphonyl side-group, it is almost totally insoluble in water (only 0.04% w/v) and must first be dissolved in acetone or ethanol.

The standard approach is to prepare a stock solution in the solvent, usually 100 $g.l^{-1}$, which, if kept in a dark, stoppered bottle, will keep for long periods (at least a year). It is easier and less wasteful to dispense small volumes of this stock than to weigh out small quantities of solids. Benzocaine hydrochloride is more soluble in water, but is more costly and forms an acidic solution. In solution, benzocaine is neutral, and probably because of this causes less hyperactivity and initial stressful reaction than MS222. However, it should be noted that buffered MS222 is almost identical to benzocaine, both chemically and in terms of physiological reactions. Although many of the side effects of MS222 are still present with benzocaine anaesthesia, the long-term consequences of these side effects do not seem to impair function. Regular anaesthesia of fish in trials at Stirling has not shown any decrement of growth or reproductive capacity.

Benzocaine is effective at approximately the same doses as MS222, 25 to 50 $mg.l^{-1}$, and is useful in freshwater, marine and tropical species, again at higher doses in the latter. Gilderhus (1989a,b) induced anaesthesia in *Oncorhynchus tshawytscha, O.mykiss* and *Salmo salar* after 3.5 minutes at 25-45 $mg.l^{-1}$, with recovery after 10 minutes following a 15 minute exposure. Salmonids generally require about 40 $mg.l^{-1}$ (Laird and Oswald, 1975) and tilapias about 100 $mg.l^{-1}$ (Ross and Geddes, 1979).

Benzocaine has a good margin of safety, although this appears to reduce at higher temperatures. Its efficacy is not affected by water hardness or pH. As with MS222, it is fat-soluble and recovery times can be prolonged in older or gravid animals.

Benzocaine is not toxic to humans at the concentrations used and is widely available in many proprietary medical preparations for human use.

Allen (1988) showed that, following exposure to 50 $mg.l^{-1}$ benzocaine for 15 minutes, tissue levels in rainbow trout, *Oncorhynchus mykiss,* and largemouth bass, *Micropterus salmoides*, were 14 $\mu g.g^{-1}$ and 10 $\mu g.g^{-1}$, respectively. Levels fell to less than the control values after 8 hours in *Micropterus salmoides* and after only 4 hours in *Oncorhynchus mykiss*, when held in running water. Generally, tissue levels of the drug fall to undetectable levels within approximately 24 hours. However, the withdrawal time required by the FDA is, again, 21 days. Fishmeal prepared from anaesthetised animals contained approximately 45 $\mu g.g^{-1}$ benzocaine (Allen, 1988).

Howe *et al.* (1990) described practical techniques for absorption of benzocaine from farm effluents using activated carbon filtration. This approach may be a consideration in obtaining approval for all farm drugs in the future.

Quinaldine (2-4-methylquinoline) and quinaldine sulphate

Quinaldine is a yellowish oily liquid with limited water solubility and it must first be dissolved in acetone or alcohol before mixing with water. Solutions in water may turn red-brown after air exposure. While it is an effective anaesthetic, it is unpleasant, irritant and insoluble and corneal damage has been reported following its use with salmonids (R.H. Richards, personal communication). The low cost of quinaldine has made it a popular tool for collection of fish from tidal pools and small lagoons. Quinaldine sulphate is a pale-yellow water-soluble powder and, although it was formerly not readily available commercially (Blasiola, 1976) it is now marketed by a number of companies. It is more pleasant to handle, but is more costly than quinaldine and MS222.

Quinaldine solutions are acidic and are usually buffered with bicarbonate. Stock solutions, usually of 10 $g.l^{-1}$, can be stored in dark, stoppered bottles.

Induction takes 1 to 4 minutes and recovery is usually rapid and uneventful. Drug susceptibility varies, but quinaldine sulphate induction solutions are in the range 15 to 60 ppm. Schoettger and Steucke (1970) noted that some reflex responsiveness was retained during surgical anaesthesia in rainbow trout and Schramm and Black (1984) found that quinaldine induction solutions were unsuitable for maintenance during surgery at 29°C in grass carp, *Ctenopharyngodon idella*. Chellappa *et al.* (1996) found that the relatively high dose of 150 ppm quinaldine was effective in facilitating handling of 7.5 to 10 kg *Colossoma macropomum* during spawning.

In general, the potency of quinaldine is less in soft water and more in warm water. Quinaldine is ineffective at pH 5 and below and is more potent at a higher pH.

Although irritant and having an unpleasant odour, quinaldine is not known to be carcinogenic.

Tissue levels of the drug fall to undetectable limits within 24 hours.

2-Phenoxyethanol (phenoxethol)

This is a clear white or straw coloured oily liquid with a slight aromatic odour which fairly easily passes into solution if shaken with a small quantity of water. The solution is bactericidal and fungicidal and because of this additional feature it is useful during laparotomy or abdominal surgery. It has no great advantages over other drugs, but is relatively inexpensive.

The drug remains effective in working solution for at least 3 days.

Doses of 0.5 $cm^3.l^{-1}$ (equal to 385 $mg.l^{-1}$) produce surgical anaesthesia in rainbow trout, *Oncorhynchus mykiss*, and sedation can be produced at a lower dose, although analgesia is sometimes incomplete. Barton and Helfrich (1981) considered that juvenile salmonids should not be exposed to this dose level, about half the LD_{50}, for more than 13 minutes and that a general dose level of 0.25 $cm^3.l^{-1}$ was preferable. Takashima *et al.* (1983) showed that yearling rainbow trout of 100 to 200 g body weight could be held for long periods at 200 ppm. They also showed that serum cortisol levels increased rapidly after immersion in the drug and concluded that stress was not alleviated in 2-phenoxyethanol anaesthesia. Sehdev *et al.* (1963) found that, at 11°C, the effective dose and lethal dose were 0.09 and 0.29 $cm^3.l^{-1}$ in sockeye salmon, *Onchorhynchus nerka*, giving a therapeutic index of over 3. The potency of the drug increased at 4°C, giving an increased therapeutic index of 5.

McCarter (1992) achieved sedation of broodstock silver carp, *Hypophthalmichthys molitrix*, and grass carp, *Ctenopharyngodon idella*, in a 0.2 $cm^3.l^{-1}$ solution and this dose did not impair sperm motility. Yamamitsu and Itazawa (1988) induced deep sedation and deep anaesthesia in adult *Cyprinus carpio* at 400 to 800 ppm, respectively, and Josa *et al.* (1992) showed that a range of effects from light sedation to surgical anaesthesia could be induced in carp at doses between 100 and 600 $cm^3.l^{-1}$.

2-Phenoxyethanol caused marked reduction in dorsal aortic blood pressure in rainbow trout (Fredricks *et al.*, 1993). Recovery is abrupt on occasions (R.L. Oswald, personal communication).

Imamura-Kojima *et al.* (1987) showed that 2-phenoxyethanol was distributed principally in the brain, liver, kidney and gall bladder of rainbow trout and that its biological half-life was approximately 30 minutes.

Metomidate (methoxymol, Marinil, Hypnodil, R7315)

Since the first edition of this handbook, a substantial amount of work has been carried out on the two imidazole-based non-barbiturate hypnotic drugs metomidate and etomidate, both of which have been used in human medicine. Both drugs produce anaesthesia without the accompanying elevation of blood cortisol. This feature was seized upon in the early 1980s and it was thought that these materials would produce stress-free anaesthesia. Kreiberg (1992) noted the relatively small increase in plasma cortisol in Pacific salmon on handling following metomidate anaesthesia. It has since been shown that this effect is achieved by the suppression of parts of biochemical pathways leading to cortisol synthesis. This is clearly undesirable and the drug etomidate has now been withdrawn from use in human medicine for the same reason.

The use of metomidate has been explored by fish biologists and it has been shown to be very effective. Induction is rapid, taking 1 to 2 minutes, and recovery is faster than after MS222 anaesthesia. Matson and Riple (1989) considered that metomidate is a better choice than benzocaine, MS222, chlorbutanol and 2-phenoxyethanol for cod, *Gadus morhua*, because of this shorter recovery time, although it can be accompanied by muscle twitching. It produces anaesthesia in salmonids at doses of only 5 to 6 $mg.l^{-1}$ and even lower doses are effective in catfish. Marine tropicals and sharks are anaesthetised at 2.5 to 5 $mg.l^{-1}$. Using 0.5 and 1.0 $mg.l^{-1}$ concentrations of metomidate, Ross *et al.* (1993), working with shad, *Alosa sapidissima,* induced sedation after 9 and 3 minutes with mean recovery times of 6 and 7 minutes, respectively. For fish exposed to 1.0 $mg.l^{-1}$ metomidate, normal swimming behaviour was delayed for as long as 4 hours after fish were placed in drug-free water and schooling behaviour was also disturbed for up to 24 hours. Olsen *et al.* (1995) induced anaesthesia in Atlantic salmon, *Salmo salar,* at a range of concentrations from 1 to 5 $mg.l^{-1}$ and noted that the drug was more potent in larger, sea water adapted fish than in freshwater parr.

By contrast with these positive indications, Massee *et al.* (1995) found doses of 3 to 4.5 $mg.l^{-1}$ to be unreliable with groups of larvae of goldfish, *Carassius auratus*, and red drum, *Sciaenops ocellatus*, giving a low percentage of anaesthetised animals with a high subsequent mortality.

Metomidate has been found to cause drastic reduction in dorsal aortic blood pressure in rainbow trout (Fredricks *et al.,* 1993). Darkening of some species has been reported and this is thought to be due to interference with hormonal control of MSH synthesis. Cortisol synthesis is inhibited during metomidate anaesthesia and Olsen *et al.* (1995) have shown that the cortisol response to handling stress was removed in Atlantic salmon, *Salmo salar,* and that blood lactate and haematocrit also increased.

Etomidate (Hypnomidate, Amidate)

This non-barbiturate hypnotic has been used intravenously for anaesthesia in humans and laboratory animals. Amend *et al.* (1982) noted effective dose rates of 2.0-4.0 $mg.l^{-1}$ in the aquarium fish *Danio rerio, Gymnocorymbus ternetzi, Pterophyllum scalare and Xiphophorus maculatus,* induction occurring within 90 seconds and recovery within 40 minutes. They suggest that dose rates up to 20 $mg.l^{-1}$ can be tolerated. Limsuwan *et al.* (1983) induced anaesthesia within 15 minutes using 3 $mg.l^{-1}$ etomidate in channel catfish, *Ictalurus punctatus*, and golden shiner, *Notemigonus crysoleucas,* although Plumb *et al.* (1983) noted that a working concentration of 0.8 to 1.2 $mg.l^{-1}$ was sufficient. Sedation was obtained at dose rates of 0.2 to 0.6 $mg.l^{-1}$ and continuous anaesthesia was maintained for 96 hours with 0.6 $mg.l^{-1}$ etomidate. They found it to give slower induction and recovery and to be more toxic at lower temperatures. The therapeutic index ranged from 4 to 7 and reduced with exposure time. Falls *et al.* (1988), working with red drum, *Sciaenops ocellatus,* found that at 0.8 $mg.l^{-1}$ induction and recovery times were excessively long. They induced anaesthesia at 1.6 $mg.l^{-1}$ and were able to maintain anaesthesia at 0.4 $mg.l^{-1}$.

Less widely used drugs for inhalation anaesthesia

It is almost certainly the case that the popular drugs already discussed retain their dominance due to the large body of literature in which they are cited or used. The following wide range of drugs has been used less frequently. In some cases the drugs are quite effective, but have simply not been widely adopted by fish biologists or veterinarians. In many cases their usefulness is somewhat limited by the lack of knowledge of their precise physiological effects, while others are no longer popular or have unwanted side effects. Dose rates for these drugs for a range of fish species are summarised in Table 10.

Table 9. Summary of dose rates for a number of fish species using the major inhalational anaesthetic drugs, MS222, benzocaine, quinaldine, 2-phenoxyethanol, metomidate and etomidate.

Drug	Species	Dose	Author
MS222	*Salmo salar*	100 mg.l^{-1}	Soivio *et al.* (1974)
	Onchorhynchus mykiss	100 mg.l^{-1} 80 mg.l^{-1}	Soivio *et al.* (1977) Wedermeyer (1969)
	Salvelinus fontinalis	100 mg.l^{-1}	Houston *et al.* (1971)
	Oncorhynchus sp.	50 mg.l^{-1}	Strange and Schreck (1978)
	Cyprinus carpio - carp fry	20-85 mg.l^{-1} 100 mg.l^{-1} 250-350 mg.l^{-1}	Takeda *et al.* (1987) Houston *et al.* (1973) Jain (1987)
	Tinca tinca	25-200 mg.l^{-1}	Randall (1962)
	Ctenopharyngodon idella	75 mg.l^{-1}	Schramm and Black (1984)
	Tilapia adults - tilapia fry	100-200 mg.l^{-1} 60-70 mg.l^{-1}	Ross and Geddes (1979)
	Etheostoma fonticola	60 mg.l^{-1}	Brandt *et al.* (1993)
	Anguilla rostrata	100-230 mg.l^{-1}	Prieto *et al.* (1976)
	Mugil cephalus	20-120 mg.l^{-1} 75-100 mg.l^{-1}	Sylvester (1975) Dick (1975)
	Polyodon spathula	66 mg.l^{-1}	Ottinger *et al.* (1992)
	Morone saxatilis	110-123 mg.l^{-1} 150 mg.l^{-1}	Henderson *et al.* (1992) Lemm (1993)
	Gadus morhua	75 mg.l^{-1}	Matson and Riple (1989)
	Pagrus major	50-100 mg.l^{-1}	Ishioka (1984)
	Pagrus auratus	60-100 mg.l^{-1}	Ryan (1992)
	Sparus aurata	70 mg.l^{-1}	Chatain and Corrao (1992)
	Dicentrarchus labrax	70 mg.l^{-1}	Chatain and Corrao (1992)
	Hippoglosus hippoglosus	250 mg.l^{-1}	Malmstroem *et al.* (1993)
Benzocaine	*Oncorhynchus mykiss*	30-50 mg.l^{-1}	Oswald (1978)
	Salmo trutta	40 mg.l^{-1}	Oswald (1978)
	Salmo salar (smolts)	40 mg.l^{-1}	Ross and Ross (1984)
	Oncorhynchus tshawytscha	25-30 mg.l^{-1}	Gilderhus (1990)
	Esox lucius	200 mg.l^{-1}	Webster (1983)
	Tilapia sp.	100 mg.l^{-1}	Ross and Geddes (1979)
	Clarias batrachus	100 mg.l^{-1} (+)	Ross and Geddes (1979)
	Prochilodus lineatus	100-200 mg.l^{-1}	Parma de Croux (1990)
	Gadus morhua	40 mg.l^{-1} 40 mg.l^{-1}	Ross and Ross (1984) Matson and Riple (1989)
	Pollachius virens	40 mg.l^{-1}	Ross and Ross (1984)
	Morone saxatilis	55-80 mg.l^{-1} 55-100 mg.l^{-1}	Gilderhus *et al.* (1991) Lemm (1993)
Quinaldine	Various spp.	10-30 ppm 10-30 ppm	Randall and Hoar (1971) Tytler and Hawkins (1981)
	Cyprinids	12-37 ppm	Osanz-Castan *et al.* (19930
	Clarius gariepinus	6 ppm	Hocutt (1989)
	Epalzeorhynchus bicolor	2-4 ppm	Meenakan and Laohavisuti (1993)
	Colossoma macropomum	100-150 ppm	Chellappa *et al.* (1996)
	Large freshwater spp.	100-150 ppm	Kaneko (1982)
	Ctenpharygodon idella	10-50 ppm	Schramm and Black (1984)
	Morone saxatilis	25-40 mg.l^{-1}	Lemm (1993)
	Lya dussumeri	100 ppm	Sylvester (1975)
	Blennius pholis	2.5-20 ppm	Dixon and Milton (1978)
	Tropical marines	200 ppm	Blasiola (1976)

2-Phenoxyethanol (2-PE)	*Oncorhynchus mykiss*	200 ppm 0.5 $cm^3.l^{-1}$	Takashima *et al* (1983) Hilton and Dixon (1982)
	Oncorhynchus nerka	0.1-0.5 $cm^3.l^{-1}$	Klontz and Smith (1968)
	Salmonids	0.25 $cm^3.l^{-1}$	Barton and Helfrich (1981)
	Cyprinus carpio	400-600 ppm	Yamamitsu and Itazawa (1988)
	Cyprinids	0.1-0.5 $cm^3.l^{-1}$	Osanz-Castan *et al.* (19930
	Hypophthalmichthys molitrix	0.2 $cm^3.l^{-1}$	McCarter (1992)
	Ctenopharyngodon idella	0.2 $cm^3.l^{-1}$	McCarter (1992)
	Epalzeorhynchus bicolor	75-200 ppm	Meenakan and Laohavisuti (1993)
Metomidate	*Gadus morhua*	5 $mg.l^{-1}$	Matson and Riple (1989)
	Salmonids	5-6 $mg.l^{-1}$	Anon
	Salmo salar	1-5 $mg.l^{-1}$	Olsen *et al.* (1995)
	Morone saxatilis	7.5-10 $mg.l^{-1}$	Lemm (1993)
	Hippoglosus hippoglosus	10 $mg.l^{-1}$	Malmstroem *et al.* (1993)
	Marine fish	2.5-5 $mg.l^{-1}$	Anon
	Alosa sapidissima	0.5-1 $mg.l^{-1}$	Ross *et al.* (1993)
Etomidate	Aquarium fish	2-4 $mg.l^{-1}$	Amend *et al.* (1982)
	Ictalurus punctatus	3 $mg.l^{-1}$ 0.9-1.2 $mg.l^{-1}$	Limuswan *et al.* (1983) Plumb *et al.* (1983)
	Sciaenops ocellatus	0.4-1.8 $mg.l^{-1}$	Falls *et al.* (1988)

2-Amino-4-phenylthiazole (APT, phenthiazamine, phenthiazamine hydrobromide, Piscaine)

This material has been thoroughly investigated for use in fish principally by workers in Japan and a good deal of work has been done on its metabolism in fish tissues. Sekizawa *et al.* (1971) described probable mechanisms for its action and suggested that it acted by depression of the CNS rather than acting peripherally. Carp, *Cyprinus carpio,* were sedated by immersion in a 12 ppm solution and this could be extended for up to 72 hours (Kikuchi *et al.,* 1974). Immersion in solutions of 30 to 40 ppm gave good anaesthesia for 20 to 40 minutes. Rainbow trout were sedated for up to 24 hours using a 10 ppm solution and were anaesthetised for 40 minutes to 3 hours using a 20 to 30 ppm solution. The marine yellowtail, *Seriola quinqueradiata*, was sedated at 8 ppm for up to 4.5 hours and anaesthetised at 15 to 20 ppm for 10 to 25 minutes. Chiba and Chichibu (1992) successfully induced anaesthesia in the loach, *Cobitis biwae,* in 7 minutes using a 50 ppm solution.

4-Styrylpyridine

This is a white powder, easily soluble in water. It is effective in a range of species at 20 to 50 $mg.l^{-1}$, inducing anaesthesia in 1 to 5 minutes and with a 20 to 30 minute recovery period (Klontz and Smith, 1969). It has also been used in the lamprey, *Petromyzon marinus,* by Piavis and Piavis (1978). It is considered safe for handling by operators.

Amylobarbitone (amobarbital, Amytal, sodium amylobarbitone)

This barbiturate is very soluble in water and is effective in salmonids at 7 to 10 $mg.l^{-1}$ (McFarland and Klontz, 1969). Induction requires 30 to 60 minutes, maintenance is good over long periods and recovery is slow, requiring up to 5 hours. It is antagonised by high calcium levels found in hard fresh waters and sea water. It is consequently not of great practical use.

Chloral hydrate

This is available as strong-smelling, colourless crystals which are freely soluble in water. It is a good CNS depressant but appears to provide poor analgesia. Surgical anaesthesia is only attained at high doses, at which problems with ventilation may occur, or heart action may be arrested. It is effective in fish as a 1% solution, slowly inducing anaesthesia in cyprinids after 25 minutes. Respiratory arrest is likely and it is not recommended to sustain immersion for longer than 10 minutes. Recovery may take up to 2 to 3 hours. Its main use has been as a sedative in transportation.

Chlorbutanol (chloretone)

This is available as crystals which dissolve easily in hot water but it is slow to dissolve in cold water. Ten per cent aqueous stock solutions remain stable for long periods and it is effective at 8 to 10 $mg.l^{-1}$ (McFarland and Klontz, 1969). Induction is rapid (2 to 3 minutes) maintenance is stable and recovery requires 30 to 60 minutes, although the anaesthetic response has been found to be very variable. It can also be prepared for use as a 30% solution in ethanol, which is then added to water at a rate of 1 $cm^3.l^{-1}$. It is effective at this dose rate for routine farm operations with Atlantic salmon (Nordmo, 1991).

Chloroform

This is a clear colourless fluid with a sweet pungent odour. It is not very soluble in water (0.8% v/v) but it can be an effective agent. However, it is now known to be very hepatotoxic to operators and should not be used in confined spaces, if at all.

Clove oil

Clove oil has been used as a mild anaesthetic since antiquity and its effectiveness as an anaesthetic in dentistry is well known. The major constituent (70 to 90% by weight) is the oil eugenol, but clove oil also contains a very wide range of turpenoid compounds which impart its characteristic odour and flavour. There are many examples of its use as an animal anaesthetic and Endo *et al.* (1972) showed that it is effective in Crucian carp, *Carassius carassius*, and Hisaka *et al.* (1986) showed that it gave effective anaesthesia in common carp (*Cyprinus carpio*) at 25 to 100 ppm. More recently, Soto (1995) found doses of 100 $mg.l^{-1}$ to be effective in the rabbitfish, *Siganus lineatus.* The fish lost equilibrium after 30 to 45 seconds and recovery required about 3 minutes. In a detailed series of experiments with juvenile (20 g) rainbow trout (*Oncorhynchus mykiss*), Keene *et al.* (1998) showed the 8 to96 hour LC_{50} of eugenol to be about 9 ppm. Doses as low as 2 to 5 ppm produced sedation sufficient for transportation work while doses of 40 to 60 ppm for 3 to 6 minutes gave effective surgical anaesthesia. In all cases recovery was dose and time related, increasing exponentially with exposure time. With short exposure times, recovery was uneventful but was always more lengthy than with MS222. Repeated anaesthesia had no detrimental effects and feeding resumed rapidly. There were no mortalities recorded and there is no apparent growth inhibition. A 10 $cm^3.l^{-1}$ (=10 $g.l^{-1}$) stock solution was found to be still effective after 3 months' storage at room temperature. The major advantage of clove oil is that is it inexpensive, not unpleasant to handle and has no known harmful effects for humans.

These features have prompted the development of a new anaesthetic compound for fish, Aqui-S, by workers at the Seafood Research Laboratory in New Zealand. The product is reported to contain 50% 2-methoxy-4-propenylphenol and 50% polysorbate 80. These materials are in the FDA's "generally regarded as safe" (GRAS) category (i.e. for food use) and require no withdrawal period. A dosage of 17 ppm seems effective for sedation of salmon and induction is stress-free. Its use is principally in harvesting of commercial fish species (mostly salmonids) where the low stress induction gives improved product quality in terms of colour, texture and external appearance. Although effective, safe to fish and humans and inexpensive, some workers consider that the very slow induction achieved at these recommended concentrations is not beneficial.

Ether (diethyl ether)

Ether is slightly soluble in water (8.4% w/w). It is a relatively safe and effective anaesthetic when used at a concentration of 10 to 50 $cm^3.l^{-1}$ in water, depending upon species, although it should be noted that, in terrestrial vertebrates it has a muscle-relaxing curare-like action. It has been used to anaesthetise a range of species, including eels, goldfish and trout. Anaesthesia is induced in 2 to 3 minutes and recovery takes 5 to 30 minutes. It has also been vaporised and bubbled through water for use with *Scyliorhinus caniculus* and *Gobius paganellus* (Vivien, 1941).

It should be borne in mind that ether is highly flammable, forms very explosive mixtures with air and must be treated with extreme caution. McFarland and Klontz (1969) note that it is highly irritant at higher working concentrations and consider that this is a strong deterrent to its use.

Lignocaine (Xylocaine, Lidocaine)

Xylocaine is widely used as a local anaesthetic and general analgesic in human medicine. It is relatively cheap, easy to obtain and likely to be safe to humans at the concentrations used with fish. This material has been used at dose levels of 100 mg.l-1 by Houston *et al.* (1973) in brook trout, *Salvelinus fontinalis*, and carp, *Cyprinus carpio*. Rodriguez and Esquivel (1995) showed that repeated Xylocaine anaesthesia in carp (*Cyprinus carpio*) had no detrimental effects. Feldman *et al.* (1975) showed that it was effective in goldfish, especially if used in conjunction with sodium bicarbonate. Rivera Lopez *et al.* (1991) obtained good anaesthesia at 150 $mg.l^{-1}$ in *Algansea lacustris*, a high-altitude Mexican cyprinid, when the drug was potentiated with sodium bicarbonate at 1 $g.l^{-1}$. Induction is uneventful and rapid, requiring 1 minute or less, and full recovery occurs after about three to four times the induction period.

Methyl pentynol (methyl parafynol, Dormisan)

This liquid has an unpleasant, acrid odour. It is soluble in water up to 147 $cm^3.l^{-1}$. It induces anaesthesia in salmonids in 2 to 3 minutes at 0.5 to 0.9 $cm^3.l^{-1}$, although animals may go into respiratory arrest during maintenance (Klontz and Smith, 1969).

Propoxate (propoxate hydrochloride, R7464)

This drug is a crystalline ester of carboxylic acid which is highly soluble in fresh water and sea water. It is said to form a stable solution for long periods. It has very impressive anaesthetic properties, being about 100 times more potent than MS222, weight for weight (Thienpont and Niemegeers, 1965). Very rapid induction occurs at high doses (30 to 60 seconds at 4 $mg.l^{-1}$). Lower doses are effective but induction takes longer (5 to 9 minutes at 1 $mg.l^{-1}$). Once anaesthetised, fish can be removed from the drug solution and will remain immobilised, if kept moist, until they are re-immersed in drug-free water.

Jirasek *et al.* (1978) noted that propoxate produced anaesthesia at only 0.25 to 0.5 $mg.l^{-1}$ in tench, *Tinca tinca*. Prihoda (1979) found that it is effective in rainbow trout at 1:500,000 (2 $mg.l^{-1}$) and that effective sedation for transportation was obtained at 1:8,000,000 (0.125 $mg.l^{-1}$). He dispensed it from a 1% stock solution. No mortalities were noted in 5 years of usage.

Propoxate appears to have a high therapeutic index. Unfortunately the drug is extremely expensive and probably because of this little is known of its metabolism or physiological effects in fish.

Quinalbarbitone (seccobarbital, Seconal)

This barbiturate is a white powder and is very soluble in water. It is used as a "short" acting hypnotic and premedicant in human medicine. It is effective as a fish anaesthetic at 35 $mg.l^{-1}$. Induction of anaesthesia is very slow, requiring 30 to 60 minutes. Maintenance is good over long periods, although recovery is very slow. Although an effective fish anaesthetic, McFarland and Klontz (1969) note that it is too slow to be useful in most cases.

Sodium cyanide

This is a blocking agent for cytochrome oxidase but it has also been widely used to harvest marine tropical fish for supply to the aquarium trade. However, Hignette (1984) notes that it may be responsible for subsequent deaths many weeks later and its use for any type of anaesthesia is not advised. The material is also very toxic to the user.

Tertiary amyl alcohol (TAA, amylene hydrate)

This is a volatile liquid which is fairly soluble in water (125 $cm^3.l^{-1}$). Randall and Hoar (1971) note that TAA gives slow induction and some hyperactivity during the early stages of recovery. It is effective within 10 to 20 minutes at 0.5 to 1.25 $cm^3.l^{-1}$ in salmonids

(McFarland and Klontz, 1969) with recovery requiring 20 to 90 minutes. Anaesthesia was induced at doses of 400 to 850 ppm (0.4 to 0.85 $cm^3.l^{-1}$) in *Epalzeorhynchos bicolor* (Mennakarn and Laohavisuti, 1993) with an LC_{50} of 1350 ppm. Alvarez-Lajonchere and Garcia-Moreno (1982) produced satisfactory deep sedation in *Mugil cephalus* postlarvae at 0.5 $cm^3.l^{-1}$ and Hilton and Dixon (1982) induced full anaesthesia in *Oncorhynchus mykiss* using 7.0 $cm^3.l^{-1}$.

Tertiary butyl alcohol (TBA)

Alvarez-Lajonchere and Garcia-Moreno (1982) produced satisfactory deep sedation in *Mugil cephalus* postlarvae at 3.5 $cm^3.l^{-1}$.

Tribromoethanol (Avertin)

This is in the form of crystals with a slight aroma which dissolve in water (25$cm^3.l^{-1}$). It is effective at 4 to 6 $mg.l^{-1}$ and induces anaesthesia in 5 to 10 minutes. Maintenance is reliable and recovery requires 20 to 40 minutes (McFarland and Klontz, 1969). The working solution decomposes on exposure to light and forms the irritant products, hydrogen bromide and dibromoacetaldehyde.

Urethane (urethan, ethyl carbamate)

Urethane is crystalline and is extremely soluble in water, giving a neutral solution. It provides a relatively shallow anaesthesia within 2 to 3 minutes, with few cardiac or respiratory effects. It also has a good therapeutic index. It is effective as a 0.5% solution in *Phoxinus phoxinus*, 2 to 4% in *Lampetra* and 1 to 5% in *Salmo trutta* (see Green, 1979), although McFarland and Klontz (1969) suggested dose rates of only 5 to 50 $mg.l^{-1}$. This drug was formerly widely used but has now been found to be carcinogenic and leucopenic (Wood, 1956). Its use has been generally discontinued and is not advised.

Inhalation anaesthesia using plant extracts

Many plants contain chemicals which have traditionally been used to harvest fish in almost all parts of the world (for examples see Jenness, 1967). Perhaps the best known are species of *Derris*, which produce rotenone, and species of *Tephrosia*, which contain tephrosin, a substance similar to rotenone. Plant extracts have also been used for predator control in ponds (Baird, 1994). Many of these extracts are extremely toxic to a wide range of animals, not just fish, whereas others are used simply as potentiators of the action of other, more effective plant extracts. The active ingredients are in many cases unknown and, with the exception of derris derivatives and some other materials with a rotenone-like action, little is understood of the mode of action or ecotoxicology of these poisons (Baird, 1994).

Some of these materials, if used in modest concentrations, have biological actions which **resemble** anaesthesia and can be exploited as such. Sinha *et al.* (1992) successfully used extracts of *Cassia fistula* to reversibly anaesthetise guppies, finding the 48 hour LC_{50} of the total alkaloid extract to be 100 mg.l^{-1}. Little guidance can be given to the use of plant extracts, and as their mode of action is poorly understood they should perhaps be employed only as a last resort.

Table 10. Summary of dose rates for a number of fish species using the minor inhalational anaesthetic drugs.

Drug	Species	Dose	Author
2-Amino-4-phenylthiazole (APT)	*Cyprinus carpio* *Oncorhynchus mykiss* *Seriola quinqueradiata*	12-40 ppm 10-30 ppm 8-20 ppm	Kikuchi *et al.* (1974)
4-Styrylpyridine	Salmonids	20-50 mg.l^{-1}	Klontz and Smith (1969)
Chlorbutanol (chloretone)	Salmonids	8-10 mg.l^{-1}	Klontz and Smith (1969)
Chloral hydrate	Fish	1%	–
Clove oil	*Siganus lineatus* *Cyprinus carpio* Salmonids - as Aqui-S *Oncorhyncus mykiss*	100 mg.l^{-1} 25-100 ppm 17 ppm 40-60 ppm	Soto (1995) Hisaka *et al.* (1985) Anon Keene *et al.* (1998)
Ether (diethyl ether)	Fish (salmonids)	10-20 cm^3.l^{-1}	Randall and Hoar (1971)
Lignocaine (Xylocaine, Lidocaine)	*Salvelinus fontinalis* *Cyprinus carpio* *Algansea lacustris*	100 mg.l^{-1} 100 mg.l^{-1} 100 mg.l^{-1}	Houston et al. (1973) Houston et al. (1973) Rivera Lopez *et al.* (1991)
Methyl pentynol (methyl parafynol, Dormisan)	Fish	0.5 -0.9 cm^3.l^{-1}	Klontz and Smith (1969)
Propoxate (R7464)	Fish (salmonids)	1-10 mg.l^{-1}	Randall and Hoar (1971)
Seccobarbital	Fish (salmonids)	35 mg.l^{-1}	McFarland and Klontz (1969)
Sodium amytal	Fish (salmonids)	7-10 mg.l^{-1}	McFarland and Klontz (1969)
Tertiary amyl alcohol (TAA)	Fish (salmonids) *Epalzeorhynchos bicolor* *Mugil cephalus*	5-6 cm^3.l^{-1} 400-850 ppm 0.5 cm^3.l^{-1}	Randall and Hoar (1971) Meenakarn and Laohavisuti (1993) Alvarez *et al.* (1982)
Tertiary butyl alcohol (TBA)	*Mugil cephalus*	3.5 cm^3.l^{-1}	Alvarez and Garcia (1982)
Tribromoethanol	Fish (salmonids)	4-6 mg.l^{-1}	McFarland and Klontz (1969)
Urethane (ethyl carbamate)	Fish (salmonids)	5-40 mg.l^{-1}	Randall and Hoar (1971)

In summary

Tables 9 and 10 show listings of effective dose rates in different species. In some cases a range of values is given which will produce different degrees of anaesthesia in different sized animals. As a guide for those new to fish anaesthesia, or who are intending to anaesthetise a new species, a selection of the preferred drug should first be made for the species closest in structure, body design and habit to the new subject. One or more suggested dose rates should then be tested on a small group of non-critical animals before proceeding with any bulk work or prolonged procedures. It is important to use a number of test animals because of probable individual variation. From this, a dose and time combination can be selected for use with a good degree of confidence.

References

Allen, J.L. (1988). Residues of benzocaine in rainbow trout, largemouth bass, and fish meal. *Progressive Fish Culturist.* 50: (1), 59-60.

Amend, D.F., Goven, B.A. and Elliot, D.G. (1982). Etomidate: Effective dosages for a new fish anesthetic. *Transactions of the American Fisheries Society.* 111 (3): 337-341.

Alvarez-Lajonchere, L. and Garcia-Moreno, B. (1982). Effects of some anaesthetics on postlarvae of *Mugil trichodon* Poey (Pisces, Mugilidae) for their transportation. *Aquaculture.* 28 (3-4): 385-390.

Baird. D. (1994). Pest control in tropical aquaculture: An ecological hazard assessment of natural and synthetic control agents. *Mitteilungen Internationale Verein Limnologie.* 24: 285-292.

Barton, B.A. and Helfrich, H. (1981). Time-dose responses of juvenile rainbow trout to 2-phenoxyethanol. *Progressive Fish Culturist.* 32 (4): 223.

Bell, G.R. (1964). A guide to the properties, characteristics and uses of some general anaesthetics for fish. *Bulletin of the Fisheries Research Board of Canada.* No. 148.

Blasiola, G.C. (1976). Quinaldine sulphate, a new anaesthetic formulation for tropical marine fish. *Journal of Fish Biology.* 10 (1): 113-120.

Bové, F.J. (1962). MS222 Sandoz. The anaesthetic and tranquilliser for fish, frogs and other cold-blooded organisms. *Sandoz News, Basle.* 13: 24pp.

Brandt, T.M., Graves, K.G., Berkhouse, C.S., Simon, T.P. and Whiteside, B.G. (1993). Laboratory spawning and rearing of the endangered fountain darter. *Progressive Fish Culturist.* 55 (3): 149-156.

Brown, L. (1987). Recirculation anaesthesia for laboratory fish. *Laboratory Animals*. 21 (3): 210-215.

Chatain, B. and Corraoa, D. (1992). A sorting method for eliminating larvae without swimbladders. *Aquaculture*. 107 (1): 81-88.

Chellappa, S., Cacho, M., Huntingford, F. and Beveridge, M. (1996). Observations on induced breeding of the Amazonian fish tambaqui, *Colossoma macropomum* using CPE and HCG treatments. *Aquaculture Research*. 27: 91-94.

Chiba, A. and Chichibu, S. (1992). High-energy phosphate metabolism in the phenthiazamine hydrobromide anesthetized loach *Cobitis biwae*. *Comparative Biochemistry and Physiology*. 102C (3): 433-437.

Dick, G.L. (1975). Some observations on the use of MS222 Sandoz with grey mullet (*Mugil chelocuvier*). *Journal of Fish Biology*. 7: 263-268.

Dixon, R.N. and Milton, P. (1978) Effects of the anaesthetic quinaldine on oxygen consumption in the intertidal teleost *Blennius pholis* (L). *Journal of Fish Biology*. 12: 359, 369.

Endo, T., Ogishima, K., Tanaka, H. and Ohshima, S. (1972). Studies on the anesthetic effect of eugenol in some fresh water fishes. *Bulletin of the Japanese Society of Scientific Fisheries*. 38: 761-767.

Falls, W.W., Vermeer, G.K. and Dennis, C.W. (1988). Evaluation of etomidate as an anesthetic for red drum, *Sciaenops ocellatus*. Red Drum Aquaculture. Proceedings of a Symposium on the Culture of Red Drum and Other Warm Water Fishes (Edited by C.R. Arnold, G.J. Holt and P.Thomas). *Contribution to Marine Science*. No. 30. 37-42.

Feldman, S., De Franco, M. and Cascella, P.J. (1975). Activity of local anaesthetic agents in the goldfish. *Journal of Pharmaceutical Science*. 64: 1713-1715.

Fredricks K.T., Gingerich, W.H. and Fater, D.C. (1993). Comparative cardiovascular effects of four fishery anesthetics in spinally transected rainbow trout, *Oncorhynchus mykiss*. *Comparative Biochemistry and Physiology*. 104C (3): 477-483.

Gilbert, P.W. and Wood, F.G. (1957). Method of anaesthetising sharks and rays safely and rapidly. *Science*. 126: 212.

Gilderhus, P.A. (1989). Efficacy of benzocaine as an anesthetic for salmonid fishes. *North American Journal of Fisheries Management*. 9 (2): 150-153.

Gilderhus, P.A. (1990). Benzocaine as a fish anesthetic: Efficacy and safety for spawning-phase salmon. *Progressive Fish Culturist*. 52 (3): 189-191.

Gilderhus, P.A., Lemm, C.A. and Woods, L.C. (1991). Benzocaine as an anesthetic for striped bass. *Progressive Fish Culturist*. 53 (2): 105-107.

Green, C.J. (1979). *Animal Anaesthesia*. Laboratory Animal Handbooks. No. 8. Laboratory Animals Ltd, London. 300pp.

Henderson-Arzapalo, A., Lemm, C., Hawkinson, J. and Keyes, P. (1992). Tricaine used to separate phase 1 striped bass with uninflated swimbladders from normal fish. *Progressive Fish Culturist.* 54 (2): 133-135.

Hignette, M. (1984). The use of cyanide to catch tropical marine fish for aquariums and its diagnosis. Comptes rendus des journees aquariologiques de l'Institut *Oceanographiques.* 10 (5): 585-591.

Hilton, J.W. and Dixon, D.G. (1982). Effect of increased liver glycogen and liver weight on liver function in rainbow trout, *Salmo gairdneri* Richardson: Recovery from anaesthesia and plasma 35 S-sulphobromopthalein clearance. *Journal of Fish Diseases.* 5 (3): 185-195.

Hisake, Y., Takase, K., Ogasawara, T. and Ogasawara, S. (1986). Anaesthesia and recovery with tricaine methanesulfonate, eugenol and thipopental sodium in the carp, *Cyprinus carpio. Japanese Journal of Veterinary Science.* 48 (2): 341-351.

Hocutt, C.H. (1989). Seasonal and diet behaviour of radio-tagged *Clarias gariepinus* in Lake Ngezi, Zimbabwe (Pisces: Clariidae). *Journal of Zoology.* 219 (2): 181-199.

Houston, A.H., Madden, J.A., Woods, R.J. and Miles, H.M. (1971). Some physiological effects of handling and tricaine methanesulphonate anaesthetization upon the Brook Trout, *Salvelinus fontalis. Journal of the Fisheries Research Board of Canada.* 28 (5): 625-633.

Houston, A.H., Czerwinski, C.L. and Woods, R.J. (1973). Cardiovascular and respiratory activity during recovery from anaesthesia and surgery in brook trout (*Salvelinus fontinalis*) and carp (*Cyprinus carpio). Journal of the Fisheries Research Board of Canada.* 30: 1705-1712.

Howe, G.E., Bills, T.D. and Marking, L.L. (1990). Removal of benzocaine from water by filtration with activated carbon. *Progressive Fish Culturist.* 52 (1): 32-35.

Imamura-Kojima, H., Takashima, F. and Yoshida, T. (1987). Absorption, distribution and excretion of 2-phenoxyethanol in Rainbow trout. *Bulletin of the Japanese Society for Scientific Fisheries.* 53 (8): 1339-1342.

Ishioka, H. (1984). Physiological and biochemical studies on the stress responses of the red sea bream, *Pagrus major* (Temminck et Schlegel). *Bulletin of the Nansei Regional Research Laboratory.* 17: 1-133.

Jain, S.M. (1987). Use of MS-222 as an anaesthetic agent for young common carp *Cyprinus carpio. Journal of Inland Fisheries Society of India.* 19 (1): 67-70.

Jenness, J. (1967). The use of plants as fish poison within the Kainji basin. In: *Fish and Fisheries of Northern Nigeria* (Edited by W. Reed). Ministry of Agriculture of Northern Nigeria. 226pp.

Jirasek, J., Adamek, Z. and Giang, P.M. (1978). The effect of administration of the anaesthetics MS 222 (Sandoz) and R 7464 (Propoxate) on oxygen consumption in the tench (*Tinca tinca* L.). *Zivocisna-Vyroba.* 23 (11): 835-840.

Josa, A., Espinosa, E., Cruz, J.I., Gil, L., Falceto, M.V. and Lozano, R. (1992). Use of 2-phenoxyethanol as an anesthetic agent in goldfish (*Cyprinus carpio*). *Research in Veterinary Science*. 53: 139.

Kaneko, K. (1982). On the removal of larger freshwater fishes. *Marine Parks and Aquaria*. 11: 39-45.

Keene, J.L., Noakes, D.L.G., Moccia, R.D. and Soto, C.G. (1998). The efficacy of clove oil as an anaesthetic for rainbow trout, *Oncorhyncus mykiss* (Walbaum). *Aquaculture Research*. 29: 89-101.

Kidd R.B. and Banks, G.D. (1990). Anesthetizing lake trout with Tricaine (MS-222) administered from a spray bottle. *Progressive Fish Culturist*. 52 (4): 272-273.

Kikuchi, T., Sekizawa, Y. and Ikeda, Y. (1974). Behavioural analyses of the central nervous system depressant activity of 2-amino-4-phenythiazole upon fishes. *Bulletin of the Japanese Society of Scientific Fisheries*. 40 (4): 325-337.

Klontz, G.W. and Smith. L.S. (1969). Methods of using fish as biological research subjects. In: *Methods of Animal Experimentation* (Edited by W.I. Gay). Academic Press, New York.

Kreiberg, H. (1992). Metomidate sedation minimises handling stress in Chinook salmon. *Bulletin of the Aquaculture Association of Canada*. 92 (3): 52-54.

Laird, L. M. and Oswald, R.L. (1975). A note on the use of benzocaine (ethyl-p-aminobenzoate) as a fish anaesthetic. *Journal of the Institute of Fisheries Management*. 6 (4): 92-94.

Limsuwan, C., Limsuwan, T., Grizzle, J.M. and Plumb, J.A. (1983). Stress response and blood characteristics of channel catfish (*Ictalurus punctatus*) after anesthesia with etomidate. *Canadian Journal of Fisheries and Aquatic Science*. 40 (12): 2105-2112.

Malmstroem, T., Salte, R., Gjoeen, H.M. and Linseth, A. (1993). A practical evaluation of metomidate and MS-222 as anaesthetics for Atlantic halibut (*Hippoglosus hippoglosus*). *Aquaculture*. 113 (4): 331-338.

Massee, K.C., Rust, M.B., Hady, R.W. and Stickney, R.R. (1995). The effectiveness of tricaine, quinaldine sulfate and metomidate as anaesthetics for larval fish. *Aquaculture*. 134 (3-4): 351-360.

Matson, N.S. and Riple, T.H. (1989). Metomidate, a better anesthetic for cod (*Gadus morhua*) in comparison with benzocaine, MS222, chlorobutanol, and phenoxyethanol. *Aquaculture*. 83 (1-2): 89-94.

McCarter, N. (1992). Sedation of grass carp and silver carp with 2-phenoxyethanol during spawning. *Progressive Fish Culturist*. 54 (4): 263-265.

McFarland, W.N. and Klontz, G. W. (1969). Anesthesia in fishes. *Federation Proceedings*. 28 (4): 1535-1540.

Meenakarn, W. and Laohavisuti, N. (1993). Some anaesthetics used in the transport of *Epalzeorhynchus bicolor* (Smith). *Proceedings of the Department of Fisheries*, Thailand.

Nordmo, P. (1991). Salmon farming in Norway. In: *Aquaculture for Veterinarians* (Edited by L. Brown). Pergamon Press, Oxford.

Olsen, Y.A, Einarsdottir, I.E. and Nilssen, K.J. (1995). Metomidate anaesthesia in Atlantic salmon, *Salmo salar*, prevents plasma cortisol increase during stress. *Aquaculture*. 134: 155-168.

Osanz-Castan, E., Esteban-Alonso, J., del Nino Jesus, A., Josa Serrano, A. and Espinoza Velasquez, E. (1993). *Proceedings of 4th National Congress in Aquaculture*. Centre for Marine Research, Spain. pp. 737-742.

Oswald, R.L. (1978). Injection anaesthesia for experimental studies in fish. *Comparative Biochemistry and Physiology*. 60C: 19-26.

Ottinger, C.A., Holloway, H.L. and Derrig, T.M. (1992). Maintenance of juvenile paddlefish as experimental animals. *Progressive Fish Culturist*. 54 (2): 121-124.

Parma-de-Croux, M.J. (1990). Benzocaine (ethyl-p-aminobenzoate) as an anaesthetic for *Prochilodus lineatus* Valenciennes (Pisces, Curimatidae). *Journal of Applied Ichthyology*. 6 (3): 189-192.

Piavis, G.W. and Piavis, M.G. (1978). Implantation of an abdominal window in the sea lamprey, *Petromyzon marinus* (Pisces: Petromyzontidae). *Copeia*. 2: 349-352.

Plumb, J.A., Schwedler, T.E. and Limsuwan, C. (1983). Experimental anesthesia of three species of freshwater fish with etomidate. *Progressive Fish Culturist*. 45 (1): 30-33.

Prieto, A., Fajer, E. and Barrera, M. (1976). Utilizacion del MS222 en la anguilla americana (*Anguilla rostrata*, Le Suer). Contribuciones del Direccion de Acuicultura. La Habana, Cuba.

Prihoda, J. (1979). Experience with use of propoxate in anaesthesia and transport of salmonids in Slovakian fishery union centres. *Biol.-Chem.-Vet.-Zivocisne.-Vyroby*. 15 (3): 283-288.

Randall, D.J. (1962). Effect of anaesthetic on the heart and respiration of a teleost fish. *Nature*. 195: 506.

Randall, D.J. and Hoar, W.S. (1971). Special Techniques. In: *Fish Physiology*. Vol 6. (Edited by W.S. Hoar and D.J. Randall). Academic Press, New York. 559pp.

Rivera-Lopez, H., Orbe-Mendoza, A. and Ross, L.G. (1991). Use of Xylocaine, potentiated with sodium bicarbonate, as an anaesthetic for fry and juveniles of Acumara, *Algansea lacustris* Steindachner 1895, from Lake Patzcuaro, Michoacan, Mexico. *Aquaculture and Fisheries Management*. 22: 15-18.

Rodriguez-Gutierrez, M. and Esquivel-Herrera, A. (1995). Evaluation of the repeated use of xylocaine as anaesthetic for the handling of breeding carp (*Cyprinus carpio*). In: The Carp. Proceedings of *Aquaculture* sponsored symposium. Budapest, Hungary, 1993. *Aquaculture*. 129 (1-4): 431-436.

Ross, L.G. and Geddes, J.A. (1979). Sedation of warm-water fish species in aquaculture research. *Aquaculture*. 16: 183-186.

Ross, L.G. and Ross, B. (1984). *Anaesthetic and Sedative Techniques for Fish*. Inst. Aquaculture, Stirling University, Stirling, 42pp.

Ross, R.M., Backman, T.W.H. and Bennett, R.M. (1993). Evaluation of the anesthetic metomidate for the handling and transport of juvenile American shad. *Progressive Fish Culturist*. 55 (4): 236-243.

Ryan, S. (1992). The dynamics of MS222 anaesthesia in a marine teleost. *Comparative Biochemistry and Physiology*. 101C (3): 593-600.

Schoettger, R.A and Steucke, E.W. (1970). Synergic mixtures of MS222 and quinaldine as anaesthetics for rainbow trout and northern pike. *Progressive Fish Culturist*. 32: 202-205.

Schramm, H.L. and Black, D.J. (1984). Anaesthesia and surgical procedures for implanting radio transmitters into grass carp. *Progressive Fish Culturist*. 46 (3): 185-190.

Sehdev, H.S., McBride, J.R. and Fagerlund, U.H.M. (1963). 2-Phenoxyethanol as a general anaesthetic for Sockeye salmon. *Journal of the Fisheries Research Board of Canada*. 20: 1435-1440.

Sekizawa, Y., Kikuchi, T. and Suzuki, A. (1971). Electrophysiological surveys on the anaesthetic properties of 2-amino-4-phenylthiazole upon carp *(Cyprinus carpio)*. *Japanese Journal of Ichthyology*. 18 (3): 128-138.

Sinha, M.J., Kumar, A., Lal, B.K., Sarkhel, B.K. and Munshi, J.D. (1992). Toxicity of alkaloid extract of *Cassia fistula* L., on common guppy, *Lebistes reticulatus* (Peters) and on the physico-chemical characteristics of water. *Journal of Freshwater Biology*. 4 (1): 7-22.

Smit, G.L., Schoonbee, H.J. and Barham, W.T. (1977). Some effects of the anaesthetic MS222 on fresh water bream. *South African Journal of Science*. 73 (11): 351-352.

Soivio, A., Malkonen, M. and Tuurala, O. (1974). Effects of asphyxia and MS222 anaesthesia on the circulation of the kidney in *S. gairdneri* Richardson. *Finnish Annals of Zoology*. 271-275.

Soivio, A., Nyholm, K. and Huhti, M. (1977). Effects of anaesthesia with MS222, neutralised MS222 and benzocaine on the blood constituents of Rainbow trout. *Journal of Fish Biology*. 10: 91-101.

Soto, C. (1995). Clove oil: A fish anaesthetic. *Western Indian Ocean Waters*. 6 (2): 2-3.

Strange, J.R. and Schreck, C.B. (1978). Anaesthetic and handling stress on survival and cortisol concentrations in yearling Chinook salmon. (*Oncorhynchus tschawytscha*). *Journal of the Fisheries Research Board of Canada*. 35: 345-349.

Sylvester, J.R. (1975). Factors influencing the efficacy of MS222 to striped mullet (*Mugil cephalus*). *Aquaculture*. 6: 163-169.

Takashima, Y., Wan, Z., Kasai, H. and Asakawa, O. (1983). Sustained anesthesia with 2-phenoxyethanol in yearling rainbow trout. *Journal of the Tokyo University of Fisheries*. 69 (2): 93-96.

Takeda, T., Yamasaki, K. and Itazawa, Y. (1987). Effect of MS222 on respiration and efficacy of forced branchial irrigation with the anesthetic solution in carp. *Bulletin of the Japanese Journal of Scientific Fisheries*. 53 (10): 1701-1709.

Thienpont, D. and Niemegeers, C.J.E. (1965). Propoxate (R7464): a new potent anaesthetic agent in cold-blooded vertebrates. *Nature*. 205: 1018-1019.

Tytler. P. and Hawkins. A.D. (1981). Vivisection, anaesthetics and minor surgery. In: *Aquarium Systems* (Edited by A.D. Hawkins). Academic Press, London.

Vivien, J.H. (1941). Contribution a l'etude de la physiologie hypophysaire dans les relations avec l'appareil genital, la tyroide et les corps supra-arenaux chez les poissons selaciens et teleosteens. *Bulletin Biologique de la France et de la Belgique*. 75: 257-309.

Webster, J. (1983). The swimbladder as a hydrostatic organ in the northern pike, *Esox luscius* (L). Ph.D. Thesis, University of Stirling.

Wedemeyer, G. (1969). Stress-induced ascorbic acid depletion and cortisol production in two salmonid fishes. *Comparative Biochemistry and Physiology*. 29: 1247-1251.

Wood, E.M. (1956). Urethane as a carcinogen. *Progressive Fish Culturist*. 18: 135.

Yamamitsu, S. and Itazawa, Y. (1988). Effects of an anesthetic 2-phenoxyethanol on the heart rate, ECG and respiration in carp. *Bulletin of the Japanese Journal of Scientific Fisheries*. 54 (10): 1737-1746.

~~~~

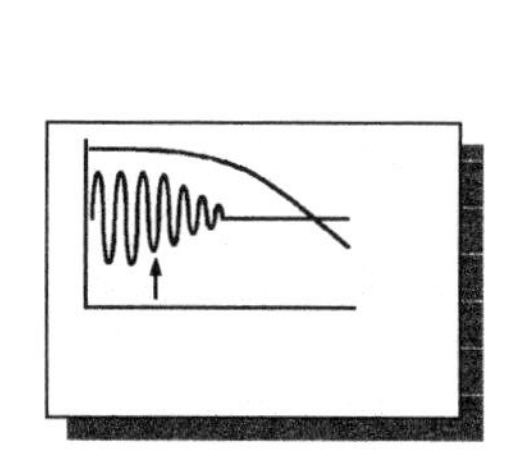
~~~~

Chapter 8
Anaesthesia of Fish
II. Inhalation Anaesthesia Using Gases

Introduction

Gaseous anaesthesia as such is virtually impossible in fish, since the fine gill lamellae collapse and fold when out of water, which increases the diffusion distance and decreases the exchange area so much as to make them virtually ineffective (Alexander, 1967). Only in fish which are capable of aerial respiration, such as catfish which have modified brush-like lamellae which do not collapse in air, one-gilled eels, the Symbranchidae, and fish with modified lungs could gaseous anaesthesia be considered feasible. It should be noted however, that even in these cases it has not been tried or tested. Consequently, only gases which are relatively soluble in water can be considered for most fish.

Carbon dioxide

The anaesthetic properties of carbon dioxide, CO_2, are well documented and it has been used as an anaesthetic with almost every animal phylum for many years. It is extremely soluble in water (880 $cm^3.l^{-1}$) and the technique simply involves bubbling the gas into the medium. It is to some extent inconvenient in that appropriate cylinders and valves are required and, furthermore, the final concentration of CO_2 in the medium may be difficult to control accurately.

It is an effective anaesthetic in fish (McFarland and Klontz, 1969) but has, until recently, only seen occasional use as a sedative for transportation (Leitritz and Lewis, 1980). It is, however, currently widely used in the salmon culture industry, where it is used for bulk "bleeding" of animals to be harvested for smoking. The nature of the anaesthesia, and particularly the analgesia, produced with carbon dioxide in these industrial circumstances has been the subject of some doubt and continuing debate.

Interest in carbon dioxide as an anaesthetic, particularly for transportation, increased during the 1980s, prompting more work on its mode of action in fish. Mitsuda *et al.*

(1988) described its use and potential for anaesthesia of carp, noting that its effects compared favourably with those of MS222 and quinaldine. Takeda and Itazawa (1983) found that the method was effective but showed that, for fully controlled anaesthesia, the pO_2 (partial pressure of oxygen) needed to be maintained at a high level (400-480 mmHg) when sedating carp at a pCO_2 of 95-115 mmHg (1 standard atmosphere=760 mmHg). They concluded that the technique was somewhat impractical because both dissolved oxygen and carbon dioxide levels needed to be controlled. Yoshikawa *et al.* (1991) noted that carp, *Cyprinus carpio,* were anaesthetised to stage II or more when the pCO_2 of the water was 125 mmHg or more. Induction took 12 minutes and reduced as the pCO_2 increased to 200 mmHg. Mean recovery times were 8 to 22 minutes.

In general, carbon dioxide anaesthesia is effective in fingerling *Oncorhynchus mykiss* at 120-150 $mg.l^{-1}$ and in adults at 200-250 $mg.l^{-1}$. Erikson *et al.* (1997) used CO_2 for pre-slaughter anaestheia of *Salmo salar* and noted that rapid anaesthesia was obtained at levels of 284 mmHg at 7°C.

Iwama *et al.* (1991) showed that hyperactivity and subsequent stress experienced when using carbon dioxide anaesthesia in trout could be reduced by buffering the water with sodium bicarbonate. Mitsuda *et al.* (1988) showed that carbon dioxide anaesthesia had a greater disruptive effect on the carp ECG than MS222, producing an extended QRS complex.

There is little doubt that controlled carbon dioxide anaesthesia is more difficult to achieve than when using other drugs.

Sodium bicarbonate

When dissolved in water, sodium bicarbonate releases carbon dioxide slowly, but will do so more rapidly if the pH is acidic.

$$NaHCO_3 = NaOH + CO_2\uparrow$$

The use of this carbon dioxide as an anaesthetic is a relatively cheap technique. Booke *et al.* (1978) investigated the use of sodium bicarbonate ($NaHCO_3$) as a fish anaesthetic. They found that it was more effective when in water with a slightly low pH. Sodium bicarbonate concentrations of 640 $mg.l^{-1}$ at pH 6.5 were most effective, inducing sedation in *Oncorhynchus mykiss, Salvelinus fontinalis* and *Cyprinus carpio* within 5 minutes. Locomotion stopped and ventilation rate decreased markedly, while equilibrium was maintained. They considered that this sedative effect resulted from the slow release of CO_2 at a rate controlled by the environmental pH. However, Gopinath Nagaraj (1990) notes that the inconvenience of adjusting and maintaining the required pH in large water volumes, so as to control the rate of CO_2 release, probably makes other methods more attractive.

Carbon dioxide released from mixtures of sodium bicarbonate and sulphuric acid was advocated for short-duration anaesthesia by Fish (1942).

$$2NaHCO_3 + H_2SO_4 = Na_2SO_4 + 2H_2O + 2CO_2\uparrow$$

He noted that a concentration of 200 ppm for *Oncorhynchus nerka* at 7 to 10°C produced anaesthesia within 90 seconds of exposure to the gas. They could be safely held in this solution for 5 minutes and recovered 10 minutes after neutralisation of the solution with sodium carbonate to remove the carbon dioxide. Post (1979) also showed that carbon dioxide released from carbonic acid (H_2CO_3) produced by deliberate acidification of sodium bicarbonate solutions was a safe, convenient and easily obtainable anaesthetic for fish. Furthermore, he noted that this is not a controlled substance in the USA and may not require FDA registration for use with food organisms. He described a convenient procedure for producing carbonic acid by adding equal volumes of 6.57% (w/v) sodium bicarbonate and 3.95% (w/v) concentrated sulphuric acid (97-98%) to a known volume of water. He noted that baths containing 150 to 600 $mg.l^{-1}$ of carbonic acid were effective and that higher concentrations induce deeper anaesthesia. Prince *et al.* (1995) describe the generation of carbon dioxide from sodium bicarbonate and glacial acetic acid.

Mishra *et al* (1983) showed that carbonic acid was an effective sedative for *Labeo rohita* and that fry could be maintained for long periods with very low mortality, significantly better than unanaesthetised controls. Kumer *et al.* (1986) reported successful use of this technique with sub-adult (15 g) Indian major carps, *Catla catla, Labeo rohita* and *Mrigal calabasu* at doses up to 600 $mg.l^{-1}$. However, heavy mortality was noted with fry above 400 $mg.l^{-1}$ and the safety margin for fry was very small. Sedation was demonstrated after 5 minutes and anaesthesia after 10 minutes of exposure. At 450 $mg.l^{-1}$ the fish could be maintained in the bath for up to 4 hours with 100% survival, although some animals began to revive after this time.

Overall, these techniques appear useful, and quite inexpensive, providing that some control of pH can be maintained. Because of the sedation it can induce at lower concentrations, sodium bicarbonate is especially useful in transportation.

Fluorinated hydrocarbons

These are a range of anaesthetic gases which are used widely in human and animal medicine. The principal members of this group are halothane (=Fluothane), enflurane (=Enthrane) and methoxyflurane (=Penthrane). All are potent anaesthetics acting on the CNS and there is some evidence that these compounds work by blocking ion transport, particularly potassium, across cell membranes. The fluothanes are normally in pure liquid form and need to be vaporised in a special apparatus at the time of use. Halothane and enflurane have relatively low boiling points of 50°C and 56.5°C, respectively, while that of methoxyflurane is much higher at 105°C.

The well-known anaesthetic of this family, halothane, has been demonstrated to be effective in fish (J. Langdon, personal communication). Dose levels of 0.5-2.0 $ml.l^{-1}$ added directly to the water produce anaesthesia. Alternatively, the gas can be vaporised and then dissolved in water to achieve the desired effect. Induction is dose related and rapid, with good surgical anaesthesia being provided and excellent maintenance and rapid recovery (2-5 minutes).

The gas is, however, relatively insoluble in water and consequently the technique is difficult to control and there is also the risk that fish can receive a fatal dose of pure halothane if the liquid is added directly to the water.

Although effective, this drug and its technique of application is probably only of minor interest in aquatic animals. Halothane is also relatively expensive.

In summary

Reversible anaesthesia using gases is possible in fish, but is of relatively minor importance. There are two important exceptions to this. Carbon dioxide, released from simple chemical reactions, can be a useful, relatively low-cost alternative in isolated locations or when other drugs are unobtainable. The most significant use of gases to sedate fish is during the bleeding process at harvest of cultured salmon for smoking. The fish are temporarily anaesthetised using carbon dioxide, the gills are cut and the animals then bleed to death. The analgesia provided and the ethics of the procedure are under debate.

References

Alexander, R. Mc N. (1967). *Functional Design in Fishes.* Hutchinson and Co., London.

Booke, H.E., Hollender, B. and Lutterbie, G. (1978). Sodium bicarbonate, an inexpensive fish anesthetic for field use. *Progressive Fish Culturist.* 40 (1): 11-13.

Erikson, U., Sigholt, T. and Seland, A. (1997). Handling stress and water quality during live transportation and slaughter of Atlantic salmon (*Salmo salar*). *Aquaculture.* 149: 243-252.

Fish, F.F. (1942). The anaesthesia of fish by high carbon dioxide concentrations. *Transactions of the American Fisheries Society.* 72: 25-27.

Gopinath Nagaraj (1990). Biotechnical considerations in the handling and transport of live fishery products. In: *Handling and Processing: The Selling Point in Malaysia Fisheries.* Occasional paper No. 3. Malaysian Fisheries Society, Kuala Lumpur, Malaysia.

Iwama, G.K., Yesaki, T.Y. and Ahlborn, D. (1991). The refinement of the administration of carbon dioxide gas as a fish anaesthetic: The effects of varying the water hardness and ionic content in carbon dioxide anaesthesia. ICES Council meeting papers. ICES-CM-1991-F27. 29pp.

Kumer, D., Mishra, B.K., Dey, R.K. and Biswas, B. (1986). Observations on the efficacy of carbonic acid as anesthetic of Indian major carps. Network of Aquaculture Cent. in Asia, Bhubaneswar (India). CIFA paper NACA/WP/86/39 (NACAWP8639). 8pp.

Leitritz, E. and Lewis, R.C. (1980). Trout and salmon culture (hatchery methods). *Californian Fisheries Bulletin*. No. 164. University of California. 197pp.

McFarland, W.N. and Klontz, G.W. (1969). Anesthesia in fishes. *Federation Proceedings*. 28 (4): 1535-1540.
Mitsuda, H., Nakajima, K., Mizuno, H., Kawai, F. and Yamamoto, A. (1980). Effects of carbon dioxide on carp. *Journal of Nutritional Science and Vitaminology*. 26 (1): 99-102.

Mishra, B.K., Kumar, D. and Mishra, R. (1983). Observations on the use of carbonic acid anaesthesia in fish fry transport. *Aquaculture*. 32 (3-4): 405-408.

Mitsuda, H., Ishida, Y., Yoshikawa, H. and Ueno, S. (1988). Effects of a high concentration of CO_2 on electrocardiograms in the carp, *Cyprinus carpio*. *Comparative Biochemistry and Physiology*. 91A (4): 749-757.

Post, G. (1979). Carbonic acid anaesthesia for aquatic organisms. *Progressive Fish Culturist*. 41 (3): 142-144.

Prince, A.M.J., Low, S.E., Lissimore, T.J., Diewert, R.E. and Hinch, S.G. (1995). Sodium bicarbonate and acetic acid: An effective anesthetic for field use. *North American Journal of Fisheries Management*. 15 (1): 170-172.

Takeda, T., and Itazawa, Y. (1983). Possibility of applying anesthesia by carbon dioxide in the transportation of live fish. *Bulletin of the Japanese Society of Scientific Fisheries*. 49 (5): 725-731.

Yoshikawa, H., Yokoyama, Y., Ueno, S. and Mitsuda, H. (1991). Electro-encephalographic spectral analysis in carp, *Cyprinus carpio*, anesthetized with high concentrations of carbon dioxide. *Comparative Biochemistry and Physiology*. 98A (3-4): 437-444.

~~~~

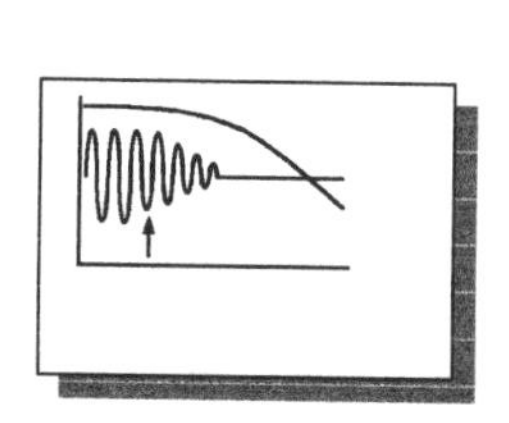
~~~~

Chapter 9
Anaesthesia of Fish
III. Parenteral and Oral Anaesthesia

Introduction

In some circumstances, particularly for surgery or lengthy procedures, anaesthesia by injection of drugs may be preferable. These techniques require slightly less specialised artificial ventilation apparatus than that required for inhalation anaesthesia, as it is not necessary to recirculate the anaesthetic solution. Usually the fish are rapidly sedated by inhalation anaesthesia to minimise handling stress, the animal is then weighed, the dose calculated and then injected. Thus, unless the initial handling of the animal can be done without anaesthesia, two drugs will usually be involved. Drugs can also be given orally, although this is less satisfactory.

Routes for administration of drugs by injection

Depending on the size of the fish, injection may be intraperitoneal, intravascular or intramuscular. The location of these access points is shown in Fig. 23.

Intraperitoneal injection is the most common and the use of a short, narrow-gauge needle ventral to the lateral line will ensure access to the peritoneal cavity. This simple method involves absorption of the drug through visceral blood vessels and induction may consequently be slow.

In larger fish, intravascular injection may be possible and, by using appropriate needles, access may be gained to certain fin sinuses, the orbital sinuses, the caudal artery and vein, the dorsal or ventral aortas or even the heart. The fin sinuses are more effective access points in larger animals. One problem with the caudal artery and vein is that, although the procedure may be successful, some damage may be done to the blood vessels in the process and this may be considered undesirable. However, repeated blood sampling by this route does not appear to cause any problems in fish of 100 g or more. The heart is not a recommended route for inexperienced workers.

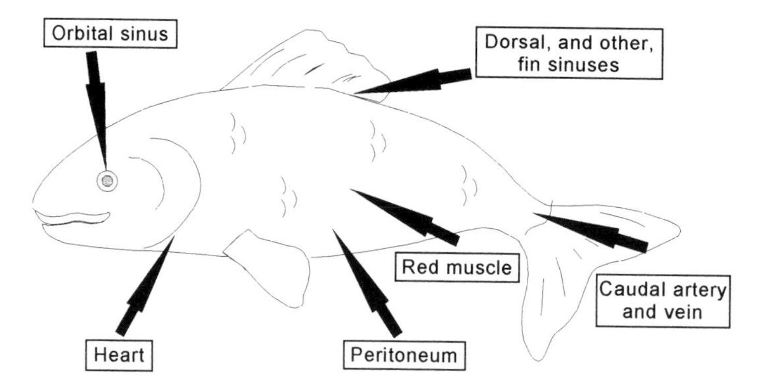

Fig. 23. Injection sites available for parenteral administration of drugs. Note that not all sites will be available in a given species.

Small drug volumes can be very rapidly absorbed if introduced into the red lateral muscle but this technique is only possible in those species having a substantial block of red muscle (Fig. 24).

The quantity of red muscle is related to the lifestyle of the fish, more active animals having more red muscle. A narrow-gauge needle should be introduced into the muscle at an oblique angle to avoid misplacement of the drug by entering the underlying mosaic, or white, muscle. The approximately 10 times greater blood flow normally found in red muscle ensures rapid movement of the drug from the site and hence relatively rapid induction. As in humans and terrestrial animals, precautions must be taken to eliminate reflux of the drug from the injection site and this is simply achieved by holding a finger over the injection point for a few seconds until the drug has been transported away from the site. Injection into the lateral musculature of fish is shown in Fig. 25.

Effective injectable anesthestic drugs

Although numerous drugs are known to be effective, there have been very few systematic investigations of injectable fish anaesthetics, the principal work having been done by Oswald (1978). Most injectable drugs used with fish originate from veterinary or human medicine; those of major interest are summarised in Table 11.

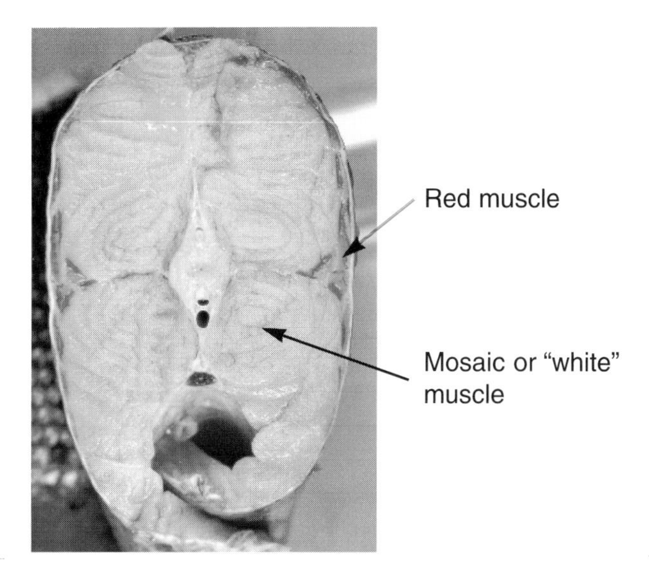

Fig. 24. The location of red muscle in Atlantic salmon (*Salmo salar*).

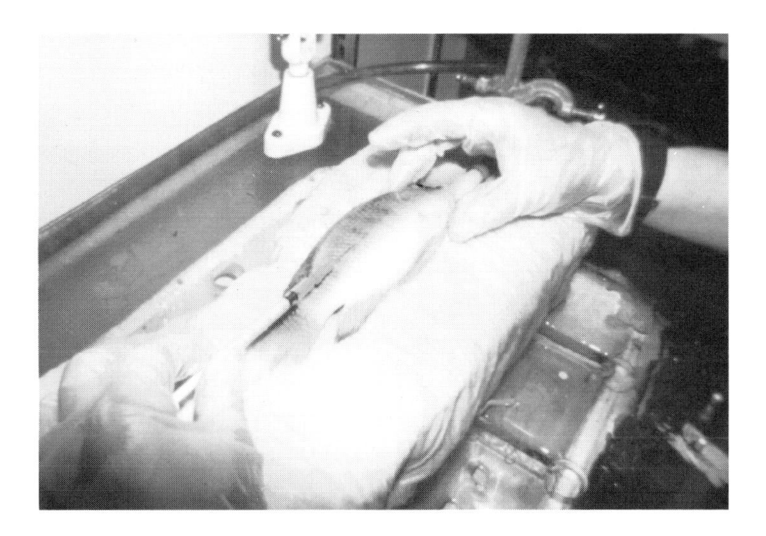

Fig. 25. Injection of the anaesthetic Saffan into the red muscle of a tilapia, prior to a lengthy procedure. Note that injection is from the rear to allow insertion of the needle under the scales.

Table 11. A compilation of drugs effective in inducing anaesthesia in fish when administered parenterally.

Name	Trade name	Characteristic
Alphaxolone-alphadolone	Saffan	Very good; long duration
Propanidid	Epontol	
Ketamine-HCl	Vetalar	
Xylazine-HCl	Rompun	
Etorphine-acetylpromazine	Immobilon	Immobilon and Revivon are used sequentially
Diprenorphine-HCl	Revivon	
Sodium pentobarbitone	Nembutal	Long duration
Sodium amylobarbitone	Amytal	
Sodium methohexitone	Brevital	Short duration
Sodium thiopentone	Pentothal	
Lignocaine-HCl	Lidocaine	Spinal anaesthesia only
Procaine-HCl		Local analgesic only

Widely used drugs for injection anaesthesia

From the range of drugs tested, effective practical anaesthesia by injection is probably limited to only a few materials. It is interesting to note that neither benzocaine nor MS222 are effective when given parenterally, probably because the rate of uptake from the site of injection is effectively lower than the gill clearance rate. The following discussion is limited to those materials known to be most widely used.

Alphaxolone-alphadolone (Saffan)

This is an excellent drug for long-term anaesthesia and in recent years has been adopted by some fish neurophysiologists because spontaneous nervous activity is essentially preserved. Its main advantage lies in its stimulatory effect upon the heart: heart beat becoming very regular and forceful. In addition, it has a general vasodilatory effect both systemically and peripherally, in contrast to the aminobenzoates (Soivio *et al.*, 1977). This effect is very obvious at the gills and this ensures adequate oxygenation of the blood.

At low doses (12 $mg.kg^{-1}$) it is possible to maintain respiration and circulation at approximately basal level in the cod, *Gadus morhua*, (Tytler and Hawkins, 1981). With higher doses (over 24 $mg.kg^{-1}$ in rainbow trout), however, ventilation becomes reciprocatory and may be abolished altogether (Oswald, 1978). Consequently, it is always advisable to anticipate ventilatory arrest. Neuromuscular preparations give consistent results over a long period using this anaesthetic. It can also be extremely valuable in studies of the cardiovascular system and, in work with saithe, *Pollachius virens*, it gave stable long-term anaesthesia with well-maintained cardiac rhythm. This resulted in steady blood flow to the gills and other organs (Ross, unpublished data) (Fig. 26).

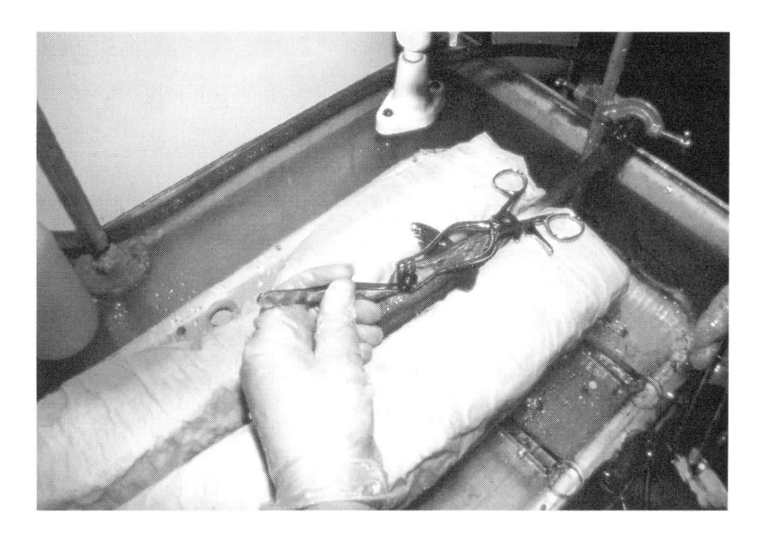

Fig. 26. Estimation of blood volume in the authors' laboratory following alphaxolone-alphadolone (Saffan) anaesthesia.

Oswald (1978) noted that there was slight excitement during recovery, but that it was essentially uneventful. Residual neuromuscular blocking effects commonly found in barbiturate anaesthesia are absent.

Ketamine (Vetalar)

Oswald (1978) found that anaesthesia persisted for only 20 minutes when rainbow trout and brown trout were injected intra-muscularly with ketamine at 130 $mg.kg^{-1}$ but sleep times of 50 to 80 minutes were obtained at 150 $mg.kg^{-1}$. Recovery time was prolonged, taking up to 90 minutes, during which there was notable hyperactivity and ataxia. Bruecker and Grahame (1993) induced anaesthesia in the cichlid *Heros citrinellum* after injection of 30 $mg.kg^{-1}$ ketamine into the dorsal aorta or caudal vein. Induction was very fast, with considerable reduction of ventilation after 1 minute. Anaesthesia lasted for up to 40 minutes and recovery required up to 4 hours. The safety margin was reduced at higher temperatures. Graham and Iwama (1990) successfully used ketamine at 30 $mg.l^{-1}$ in *Oncorhynchus mykiss* and *O. kisutch*. Smith (1992) describes successful use of ketamine in combination with xylazine (Rompun) with grey nurse sharks, *Carcharias taurus*.

Nembutal (sodium pentobarbitone, pentobarbital)

This is a barbiturate, normally of intermediate duration in mammals, which is easily available as a very stable injectable solution. Shelton and Randall (1962) anaesthetised tench, *Tinca tinca,* and roach, *Rutilus rutilus,* using intramuscular injections of 20 $mg.kg^{-1}$. It is effective in rainbow trout and brown trout at higher intraperitoneal doses of 48-72 $mg.kg^{-1}$ (Oswald, 1978). Anaesthesia is very lengthy, extending from 6-24 hours depending on dose. The main complications are that recovery is very prolonged with persistent ataxia. There is usually intense bradycardia, ventilatory arrest and the drug has some curare-like properties. By contrast, in some elasmobranchs, Nembutal has been reported to be effective at only 6 $mg.kg^{-1}$ but fatal at 60 $mg.kg^{-1}$ (Walker, 1972), thus giving a therapeutic index of about 10. Unlike most other anaesthetic agents, this drug is not excreted by the gills to any great extent, which may explain its long action. Initial recovery is probably by redistribution within the fish.

Propanidid (Epontol)

Although doses of 8-9 $mg.kg^{-1}$ propanidid are effective in producing very short anaesthesia in mammals (Clarke and Dundee, 1966), intraperitoneal doses of 325 $mg.kg^{-1}$ are required for effective anaesthesia in rainbow trout and its action lasts for about 2.5 hours (Oswald, 1978), a much longer sleep time than in mammals. Its chief advantages are that it does not greatly depress ventilation. Recovery is comparatively uneventful, with only a brief ataxic phase, and full reaction to stimulation returns after about 1 hour. Oswald (1978) considered that longer sleep times in fish were either due to the intraperitoneal injection route and associated lower clearance rates, or to the lower levels of cholinesterases found in fish plasma.

Less widely used drugs for injection anaesthesia

While the previous section describes the injectable drugs of choice for fish, some other materials have potency and are described briefly here.

Etorphine-acetylpromazine (Immobilon)

As has been found in many other cases, much higher doses of this drug were needed to produce anaesthesia in fish than in mammals. Oswald (1978) found that good anaesthesia was induced in rainbow trout at 8 to 10 $mg.kg^{-1}$. After up to 1 hour, the anaesthesia was reversible within 5 minutes by intramuscular injection of equal amounts of the specific antidote diprenorhine (Revivon). Recovery was accompanied by brief ataxia and swimming returned within 6 minutes. A notable point here is that the quantity of etorphine needed to produce anaesthesia in fish would be fatal to a human. Consequently, this drug can only be used with great care, by workers experienced in injection methods and with rapid access to the ready-prepared antidote, Revivon.

Xylazine (Rompun)

Oswald (1978) tested a range of concentrations of xylazine and found the minimum anaesthetic dose in rainbow trout to be 100 $mg.kg^{-1}$. Ventilatory collapse and convulsions occurred persistently and the ECG was grossly disturbed. Although actually effective, this drug cannot be recommended for use with fish.

Lignocaine (Xylocaine, Lidocaine)

Spinal anaesthesia induced by injection of 0.1 cm^3 lignocaine immobilised rainbow trout in 5 minutes (Oswald, 1978). There was no sensitivity posterior to the injection site and recovery required 45 to 50 minutes, with fish swimming freely after a further 15 minutes.

Procaine

Kisch (1947) produced general anaesthesia lasting for 0.5 to 1 hour by intracranial injection of procaine. The fish recovered normally.

Oral anaesthesia: chemicals in food

This method is relatively stress-free and has been used by Murai *et al.* (1979), who fed pellets containing diazepam (Valium) to American shad, *Alosa sapidissima,* at levels ranging from 0.04 to 0.08 $mg.kg^{-1}$ of fish. Although at first sight this method is very attractive, there are certain problems, which include the technicalities of incorporating the material in the diet. Furthermore, induction will tend to be slow as the drug is absorbed via the gut and it is not possible to predict accurately the quantity consumed by individuals in the tank.

Additives – again

It is important to bear in mind that many commercially available injectable preparations contain additives such as bactericides, detergents, solvents or stabilisers. It is thus necessary to exercise caution so that these components do not damage the gills where recovery is important. For example, both Epontol and Saffan contain the powerful surfactant "Cremophor EL". This is known to induce histamine release in dogs but its effect in fish is unknown.

In summary

Effective injection anaesthesia of fish is available, but has the initial disadvantage of requiring immersion or other immobilisation to facilitate the injection. Care must then be taken to calculate and administer the dose correctly, avoiding loss of material from the injection site. Overall, this method is probably only useful in the physiological laboratory.

References

Bruecker, P. and Graham, M. (1993). The effects of the anesthetic ketamine hydrochloride on oxygen consumption rates and behaviour in the fish *Heros (Cichlasoma) citrinellum* (Günther, 1864). *Comparative Biochemistry and Physiology.* 104C (1): 57-59.

Clarke, R.S.J. and Dundee, J.W. (1978). Survey of experimental and clinical pharmacology of propanidid. *Anaesthesia and Analgesia: Current Researches.* 45: 250-276.

Kisch, B. (1947). A method to immobilise fish for cardiac and other experiments with Procaine. *Biological Bulletin.* 93: 208. Marine Biological Laboratory, Woods Hole, MA.

Graham, M.S. and Iwama, G. (1990). The physiologic effects of of the anaesthetic ketamine hydrochloride on two salmonid species. *Aquaculture.* 90 (3-4): 323-331.

Murai, T., Andrews, J.W. and Muller, J.W. (1979). Fingerling American shad: effect of Valium, MS-222, and sodium chloride on handling mortality. *Progressive Fish Culturist.* 41 (1): 27-29.

Oswald, R.L. (1978). Injection anaesthesia for experimental studies in fish. *Comparative Biochemistry and Physiology.* 60C: 19-26.

Shelton, G. and Randall, D.J. (1962). The relationship between heart beat and respiration in teleost fish. *Comparative Biochemistry and Physiology.* 7: 237-250.

Smith, M.F.L. (1992). Capture and transportation of elasmobranchs, with emphasis on the grey nurse shark (*Carcharias taurus*). *Australian Journal of Marine and Freshwater Research.* Special issue: Sharks: Biology and Fisheries. 43 (1): 325-343.

Soivio, A., Nyholm, K. and Huhti, M. (1977). Effects of anaesthesia with MS222, neutralised MS222 and benzocaine on the blood constituents of Rainbow trout. *Journal of Fish Biology.* 10: 91-101.

Tytler, P. and Hawkins, A.D. (1981). Vivisection, anaesthetics and minor surgery. In: *Aquarium Systems* (Edited by A.D. Hawkins). Academic Press, London.

Walker, M.D. (1972). Physiologic and pharmacologic aspects of barbiturates in Elasmobranch. *Comparative Biochemistry and Physiology.* 42A: 213-221.

~~~~

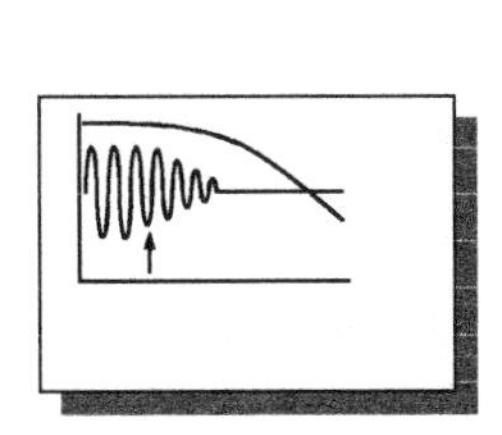
~~~~

Chapter 10
Anaesthesia of Fish
IV. Non-chemical Methods

Introduction

It is possible to immobilise fish without the use of chemicals and the two techniques available, namely hypothermia and electroanaesthesia, may have distinct advantages over chemical methods for certain procedures.

Hypothermia

Lowering the water temperature will tranquilise or immobilise fish (Bell, 1964). Lower temperatures increase the oxygen-carrying capacity of the water and reduce the activity and oxygen consumption of fish. They also reduce general metabolism and hence minimise ammonia and solid waste production. Cooling can be achieved by refrigeration or by the addition of ice, or by using dry ice in thermal contact with the water but physically and chemically isolated from it (Solomon and Hawkins, 1981). Although immobilisation is achieved and there is some reduction in sensitivity to stimulation, it must be remembered that true anaesthesia, and particularly analgesia, is not likely to be achieved in this way.

A given species of fish usually has a fairly wide temperature tolerance. However, the amount of cooling or heating which can be applied instantaneously to a fish depends on its previous temperature history and its acclimation temperature. Thus, although hypothermia can immobilise fish and lower the metabolic rate, the degree of cooling which can be applied in a practical situation may be very limited (Randall and Hoar, 1971). Gradual cooling is preferred as rapid chilling can produce a lethal shock. In addition to immobilising and depressing metabolism, low temperature is also known to cause disruption of osmoregulatory capabilities, which may result in major ion imbalances over short periods, resulting in death. Randall and Hoar (1971) remarked that "fish can be cooled to about 4°C (depending on species and thermal history) to produce a deep narcosis from which recovery is rapid on return to the acclimation temperature". It should be clear that this temperature applies only to temperate animals.

Various authors have used the technique, principally for transportation (Fish and Hanavan, 1948; Anon, 1982). Parker (1939) suggested immersion in crushed ice for 10 to 15 minutes as a technique for allowing minor procedures, and Hawkins (personal communication) found that adult Atlantic salmon could be cooled to the temperature of iced water, almost 0°C, producing a form of sedation which enabled long-distance transportation of broodstock with no attendant mortalities.

While simple hypothermia is one of the oldest fish anaesthetics, it is not generally recommended that it be used alone. Some attendant mortalities have been noted when using hypothermia and it is not clear whether these are attributable to excessive cooling or to the additive action of chemical anaesthetics, with which it is often used simultaneously. Yoshikawa *et al.* (1989) found that carp previously acclimated to 23°C could be safely maintained in a state of apparent anaesthesia for 5 hours at 4°C, whereas sedation achieved at 8 to 14°C could be maintained for 24 hours. The anaesthesia at lower temperatures produced haemorrhaging and deaths in some cases. In practice it has been found that an instantaneous temperature decrease of only 6°C can be applied to tilapia fry acclimated to 25°C, greater rates of decrease producing measurable mortalities (Okoye, 1982). When hypothermia was used in conjunction with chemical anaesthesia, the normally effective dose of benzocaine for sedation of tilapia fry (about 30 to 32 $mg.l^{-1}$) had to be reduced by about 30% to 25 $mg.l^{-1}$. Williamson and Roberts (1981) used cooling after initial MS222 anaesthesia to immobilise dogfish during long surgical procedures. No extra drug was given, electromyographic evidence for blocking of the central nervous system was obtained and eventual recovery was rapid.

Overall, this technique has good calming properties and allows handling and transportation. When used alone, however, it is not suitable for any invasive work and almost certainly gives very incomplete analgesia.

Electroanaesthesia

An interesting alternative to chemical anaesthesia is the use of electricity. Direct currents (D.C.) and alternating currents (A.C.) with sinusoidal, triangular and square waveforms are illustrated in Fig. 27 and have all been used for anaesthesia at various times. Alternating current and square waves in the form of chopped D.C. have all been used in electrofishing for many years (Vibert, 1967; Cowx and Lamarque, 1990). All four types are effective for use in fresh water, whereas only sine waves or chopped D.C. are fully effective in sea water. All are capable of producing immobilisation in fish (Scheminsky, 1934) but their modes of action differ.

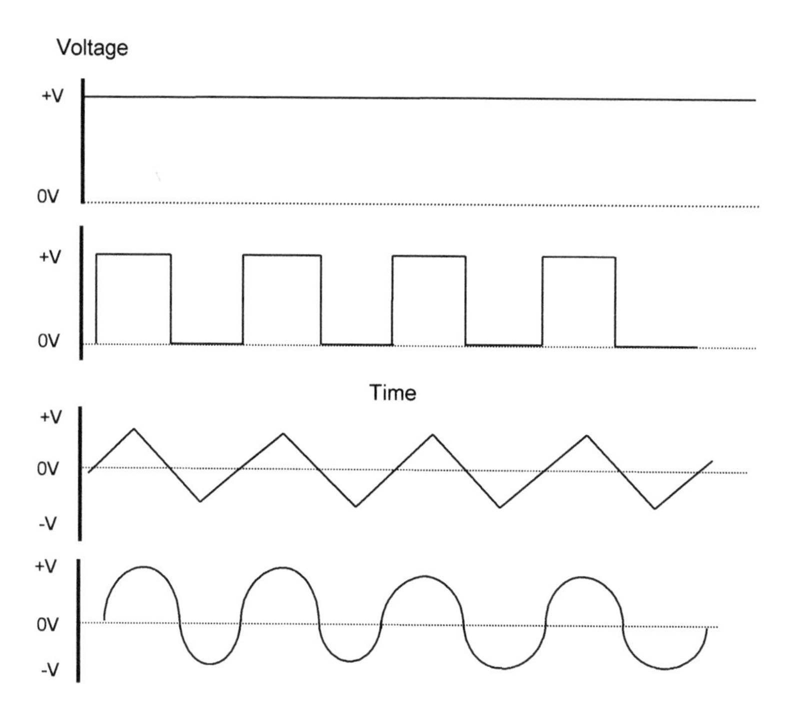

Fig. 27. Alternating and direct current waveforms. Top to bottom: direct current (D.C.), square wave (pulsed D.C.), triangular wave A.C., sine wave A.C. Note that square waves are shown here as positive pulses but that pulses can be negative or bidirectional (i.e. alternating).

Generally, fish in sea water are less susceptible to electroanaesthesia. The effect is almost certainly related to body conductivity relative to that of the water and is certainly best visualised in this way. Where the animal is less conductive than the medium (as in saline waters) electrons can flow more easily through the water directly between the electrode plates (Fig. 28a). By contrast, in fresh water the fish are more conductive than the medium and the easiest plate-to-plate route for electrons is via the fish (Fig. 28b).

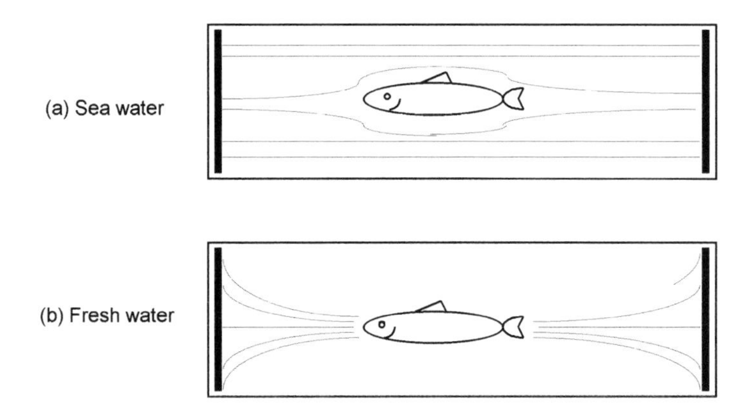

Fig. 28. Electron flow through and around fish bodies in (a) sea water and (b) fresh water.

The terms **electroanaesthesia, electronarcosis, galvonarcosis** and **electro-immobilisation** are all used in the literature, often inaccurately, and require some explanation. Electroanaesthesia and electronarcosis are often used to describe the effect produced by either A.C. or D.C., whereas galvonarcosis is usually reserved for the transitory immobilising effect produced by D.C. (see below). Both forms produce electroimmobilisation but until recently it has not been clear whether either produces real anaesthesia and some therefore consider it prudent to describe both forms as electroimmobilisation. Here, galvonarcosis is used for D.C. induced effects and electroanaesthesia is used for A.C. induced effects.

Immobilisation using direct current (D.C.)

This is usually distinguished by the term galvonarcosis (Halsband, 1967). Fish subjected to low-voltage D.C. undergo a forced swimming motion, but at higher field strengths (>300 $mV.cm^{-1}$) they undergo a simple tetany and become immobile (Lamarque, 1990). However, this is only effective while the fish are in the electrical field and after turning off the current, or if the fish escapes from the electrical field, it will recover almost immediately.

Kynard and Lonsdale (1975) demonstrated rapid paralysis of yearling rainbow trout subjected to field strengths of 0.6 $V.cm^{-1}$ D.C. The fish move initially towards the anode, lose equilibrium and become immobile in rapid succession (Halsband, 1967). By contrast, tilapia require a D.C. field strength of approximately 3 $V.cm^{-1}$ to produce a similar effect (Ross and Ross, 1984). Kolz (1989) immobilised goldfish, *Carassius auratus*, when applied currents exceeded 3 mA and calculated that the minimum electrical power required for D.C. anaesthesia was 20 mW. Recalculation of this data shows that the total intercepted (head to tail) voltage was <1 $V.cm^{-1}$ (power = current x voltage) which is quite consistent with Kynard and Lonsdale's results.

Basic galvonarcosis procedure

Power sources for galvonarcosis are usually one or more motor car batteries, appropriate cables and a simple immersed electrode pair, or array, often constructed as part of a holding box, net or basket. Electrodes should preferably be of good quality stainless steel for durability. Fish are either guided into the field in water, are netted and placed into it, or are netted in a device which contains the electrodes. The current is continuously applied and immobilisation occurs quickly, often accompanied by minor tail flicks. Having established immobility, it is often subsequently possible to reduce the field strength to a point where it is safe for the operator to handle the fish in water with little or no effect. Rubber gloves may be useful, but the D.C. field strength will usually be so low as to be undetectable by adult humans. In this condition simple procedures can be carried out so long as the fish remains with its head towards the anode and within the applied electrical field. Gunström and Bethers (1985) describe the use of a landing basket fitted with stainless electrodes powered from a 12 volt car battery. They used this apparatus to immobilise troll-caught salmon and to conduct simple, rapid procedures before turning off the current with immediate recovery. Orsi and Short (1987) describe a similar, portable device fitted with multiple electrodes and note that the smallest *Oncorhynchus tshawytscha* that could be immobilised was 20 cm in length. Larson (1995) describes the use of a padded cradle for use with larger salmonids and claims that this reduces scale loss.

D.C. galvonarcosis has been most widely applied in fisheries field work of this type. The procedure probably only immobilises and does not induce true anaesthesia; the extent of analgesia is unknown.

Immobilisation using alternating current (A.C.)

A.C. effectively produces a form of anaesthesia and the most usual waveform used is a sine wave. This is most frequently derived from the domestic mains, although a portable generator can also be used. Frequencies of 50 Hz or 60 Hz are thus common in such work. Barham *et al.* (1987b) used a signal generator and amplifier in the laboratory to study the effects of waveform and frequency. They showed that frequencies up to 200 Hz have substantially the same effect as lower, mains-derived frequencies. They also showed that square waves are just as effective as sine waves but that triangular waves are less effective. However, both square and triangular waves have the disadvantage that they need to be generated electronically, necessitating more specialist equipment, including signal generators and power amplifiers capable of delivering a high current into water.

Following A.C. treatment, the effects are not abolished by switching off the supply and a short-duration sedation or anaesthesia-like state is produced. Ludwig (1930) and Scheminsky (1934) studied the effects of alternating currents on fish and described three successive responses in fish behaviour to increasing electric current. At very low levels, the electrical stimulation results only in **electrotaxis**, but at higher voltages a form of **sedation** occurs, which is followed at even higher voltages by **electroanaesthesia**.

The reactions of fish to an electric field can differ markedly and depend upon the intensity of the electrical field, the duration of the electrical stimulation and the morphology of the fish's body (Ellis, 1975). It has been demonstrated that, assuming similar orientation, larger fish intercept a greater potential difference than small fish (Holzer, 1932; Halsband, 1967; Vibert, 1967) and consequently larger fish are affected more rapidly by relatively low potentials than small fish. A further consequence, demonstrated in *Oreochromis mossambicus* by Barham *et al.* (1987a) is that, when exposed to a set electrical stimulus, narcosis time increases markedly with body length. It has consistently been shown that the single most important factor in this is the head-to-tail voltage. It should be clear from this that a longer fish will intercept a greater voltage than a shorter one and that body shape will play an important part in determining the relative effect of a given electric current. Lamarque (1990) relates this effect, in part, to the length of nerve fibres in fish of different body length. Furthermore, the anaesthetic effect will obviously be altered when the fish are not orientated roughly parallel to the direction of electron flow. Darroux (1983) showed that a rectangular tank with full-width plate electrodes is best at producing a uniform field strength. In circular tanks fish can "escape" during the electrotaxis phase to an area of low and ineffective field strength.

Barham and Schoonbee (1990) used a fixed stimulus in attempting to induce a short handling period, and found that mature *Oreochromis niloticus* responded better to a 60 V 50 Hz A.C. stimulation than to D.C., but that half-wave rectified A.C. needed a much higher voltage of 200 V. In fact, the sedative or anaesthetic effect is graded according to the duration and magnitude of the applied voltage. Although such data are rarely available for a given species, a full dose-response diagram for tilapia is shown in Fig. 29.

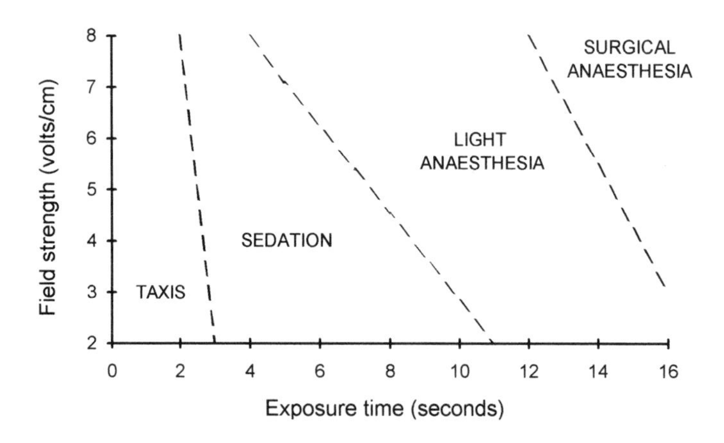

Fig. 29. Effects of electrical field strength and duration of exposure on *Oreochromis niloticus*. Note that the effects are graded from simple taxis to deep anaesthesia. Data from Robinson (1984).

Basic considerations for electroanaesthesia

The following approach has been used successfully and can easily be modified where needed to suit local circumstances.

1. The critical factor for successful electroanaesthesia is to achieve a uniform field density and some minimum size-dependent head to tail voltage. An approximately uniform field density for anaesthetic purposes can best be achieved in a rectangular tank with maximally sized, rectangular, stainless-steel electrodes placed at each end. Stainless-steel electrode plates will not corrode or shed materials into the water. Users will need to experiment with other shapes of tank, bearing in mind that field density could be made more uniform by using multiple electrodes. The maximum length of the tank must not be such as to reduce the head to tail voltage of the fish size being used to less than that required for immobilisation.
2. Arrange the tank with the electrodes firmly clamped at each end and in such a way that the fish cannot swim **behind** the electrodes. Aerate and control the temperature of the water, if needed.
3. The most convenient A.C. sources are either the domestic mains or a portable generator providing an equivalent of the domestic mains. The available voltage will thus be in the region of 240-250 V A.C., or as low as 110 V A.C. in some countries. User safety is of paramount importance and, if mains or a generator are used, the voltage should be applied to the electrodes via an isolating transformer of modest current capability, thus protecting the operator to some extent. An appropriate isolating transformer will normally have a series of secondary tappings which can be switched, thus giving some simple control over the total applied voltage. An alternative A.C. source could be a signal generator and power amplifier, the power stages of which will be current limited, depending upon the semiconductors used.
4. The unit must be operated with appropriate overload circuit breakers on the input (mains or generator) side. However, it will be obvious that such units cannot be operated with a traditional earth leakage circuit breaker, as this will cause immediate tripping and defeat the objective. Such protection devices cannot be expected to work correctly in water-filled tanks and so a resistive circuit breaker should be used. The unit should preferably have overload protection on the output side which will serve two purposes, firstly to protect the output stages from dead-

shorting, but secondly to interrupt the output should current rise above a certain limit, which also partly serves to protect the operator against accidental immersion.

5. In the simplest arrangements, it may be possible to time the application of the current for the few seconds necessary using a stopwatch, but this is much more reliably achieved using a simple, inexpensive electronic timer controlling the output relay.
6. In addition to these minimum requirements, as much protection as possible should be afforded to those in the vicinity of the operation. This can be achieved by the use of serial lock-out switches on the control box to prevent gratuitous or accidental operation and by the use of audible warning signals and flashing lights. In the authors' laboratory the unit has a four-step operating sequence of (1) selecting voltage and (2) selecting time and then (3) "arming" the output by pressing a further switch. This "arming" process illuminates warning lights, sounds a small siren and controls a relay which allows the current to be passed to the final "firing" relay. The unit must then be (4) "fired" within 20 seconds of arming, after which period it will time out and need to be fully reset again. After firing, the unit has a further protective timer which prevents re-arming within 30 to 40 seconds. This affords some short-term protection to assistants or handlers who may begin work with the animals immediately after the firing.
7. Users are reminded that no simple electrical or electronic protection can be provided that will fully protect an operator who places hands or body into the water while the current is on and hence a strict working protocol must be adhered to by all involved. Codes of practice for use of electric fishing apparatus exist in most countries and these form a useful starting point for developing a suitable protocol for electroanaesthesia (see various authors in Cowx, 1990, and Cowx and Lamarque, 1990).

Construction features for an electroanaesthesia unit

The basic building blocks are as follows:

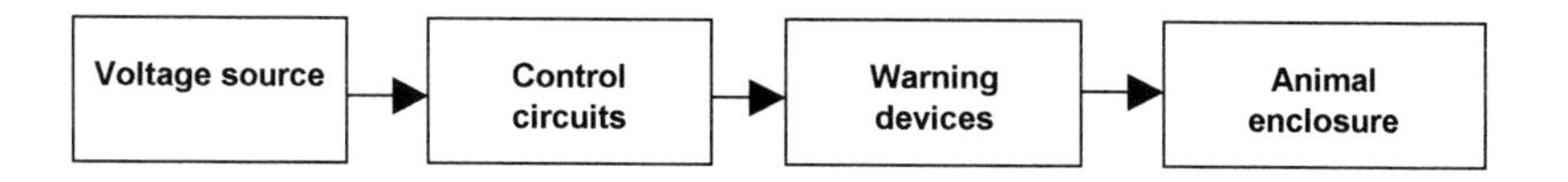

The schematic diagram in Fig. 30 shows the features of the electrical system used by the authors for controlling and delivering electroanaesthesia and Fig. 31 shows the actual unit. All of the components used in such a unit are routinely available from major electronic equipment and component distributors:

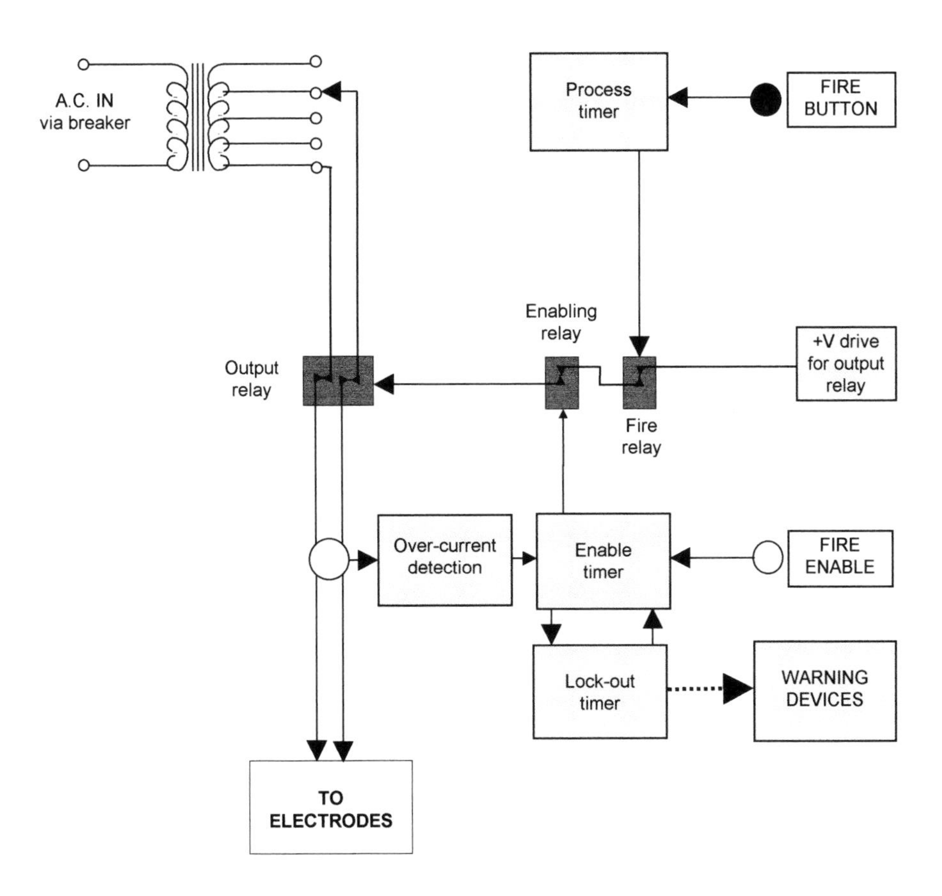

Fig. 30. Schematic diagram of electroanaesthesia apparatus.

Isolating transformer: this is a transformer of modest current capacity, typically 200 VA, which provides a range of output voltages that can be selected by an appropriate switch. Versions with a 240 volt primary will give 240 volts output, which is sufficient for most electroanaesthesia purposes. In countries with 110-120 volt supplies a step-up transformer capable of giving 240 volts output could be selected to give greater flexibility of output.

Timers: in this design, use is made of the 555 timer and the dual 556 timer in their CMOS versions. These require fairly low current and minimal additional control components. They are easily set up to give the range of times required with good precision and repeatability.

Relays: these are low-voltage types with modestly rated A.C. contacts. Although other timer circuits can be used, the 555 and 556 timers have the advantage of being capable of driving small relays of this type directly.

Switches: the fire button and fire enable button should be of a prominent push-to-make variety. It would also be sensible to have a key-operated master switch, which would prevent unauthorised persons from tampering with or attempting to use the equipment unattended. All switches should be waterproof to IP65 or better.

Warning devices: simple low-voltage sirens and bright krypton flashers are now available very cheaply and so this simple contribution to safety cannot be ruled out on the grounds of cost, although availability in some parts of the world may present difficulties. The sirens can be fitted inside the casing without loss of effect and a flasher can be mounted on top. The devices should be waterproof to IP65 or better.

Plugs and sockets: all such fittings should be of the shrouded type to avoid accidental contact of fingers with "live" pins during assembly. They should also be of high-grade, durable materials which will last well even if they become damp and should be waterproof to IP65 or better.

Case: this should be waterproof or at least splashproof, i.e. to IP65, 66 or 67, as it is almost impossible to work and remain completely dry.

The physiological effects of electroanaesthesia

Robinson (1984) investigated electroanaesthesia in tilapia, trout and eels, representing fish with three very different body forms. As may be expected, eels intercept high voltages even in weak fields and tetanise, often grotesquely, with some attendant haemorrhaging. Anaesthesia is not effective at these low field densities and so eels, or indeed any other fish with this body shape, are quite unsuited to the technique. Barham *et al.* (1989) noted muscular spasm and haemorrhaging in electrically anaesthetised carp and suggested that this species was also an unsuitable candidate for the procedure.

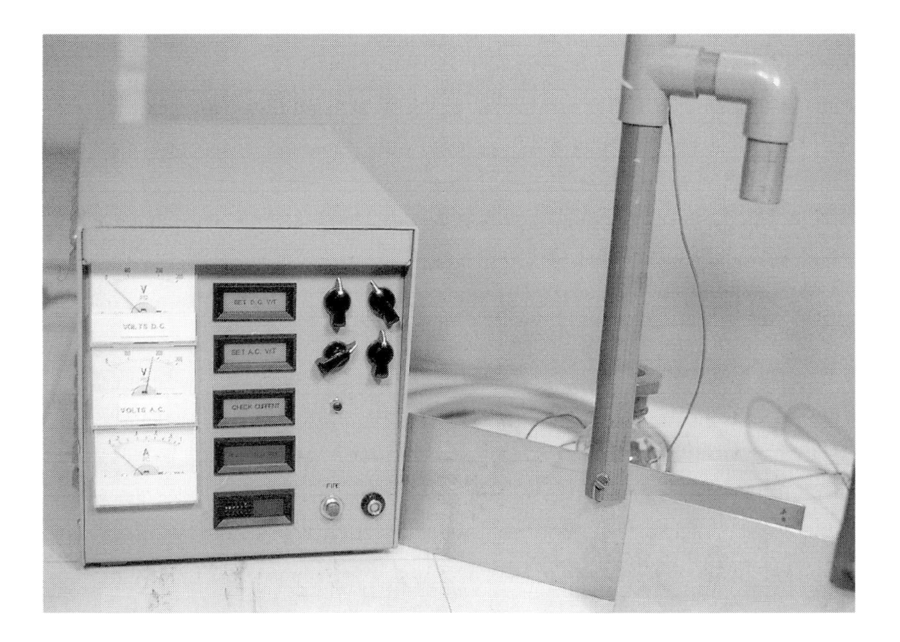

Fig. 31. The electroanaesthesia control unit and stainless steel plate electrodes used in the authors' laboratory. Note the voltage and time selector switches at top right, the indicator panels which light in sequence to guide the user through the firing procedure and the stainless-steel plate electrodes used in tanks.

Robinson (1984) showed that exposure to any voltage produced brief opercular flaring in rainbow trout and tilapia, *Oreochromis niloticus*. At voltages which induced electroanaesthesia, this was followed by decreases in ventilation rate of 50 to 70%. In some cases ventilation ceased altogether, to be resumed after a brief delay at a much lower rate. Generally, trout became paler and more silvery in appearance, whereas tilapia blanched at first but darkened in the period following the exposure. Black bands frequently appeared on the flank of both species and, although these usually disappeared within 1 hour, in a few cases they were retained permanently.

Electroimmobilisation produces haematological changes broadly similar to those from chemical anaesthesia (Madden and Houston, 1976; Schreck *et al.*, 1976). Indeed, blood electrolytes and body water distribution may undergo large changes (Schreck *et al.*,1976) and these may last for a day or more. Barham and Schoonbee (1990, 1991a,b) noted that A.C. electroanaesthesia produced changes in blood parameters which were less than, or broadly similar to, those induced by benzocaine, depending upon the parameter, but that D.C. galvonarcosis disrupted blood parameters more markedly. Robinson (1984) showed that electroanaesthetised rainbow trout and tilapia, *Oreochromis niloticus*, experienced haemoconcentration and an overall loss of water to the environment.

The great advantage of electroanaesthesia is reduced netting stress both of the fish and of the operator, as rapid immobilisation of large numbers of animals can be achieved even in relatively large holding facilities. As the head-to-tail voltage is important, it will be clear that tanks over a few metres in length will require prohibitively large voltage sources to achieve anaesthesia. The method may also have some advantages in biochemical or nutritional studies where the fish may need to be immobilised instantly. Further it is likely that, if this technique could be fully refined, it would be the method of choice for air-breathing species.

Analgesia during A.C. electroanaesthesia

In the only reported study of the analgesia produced during A.C. electroanaesthesia to date, Robinson (1984) showed that loss of reactivity to simple touch stimuli and to topical pain stimuli, judged by contact with a pencil point, lasted for less than 1 minute at lower voltages. However, this could be extended to up to 5 minutes at higher voltage and time combinations.

Pulsed white noise

Numerous authors have noted that electrically anaesthetised fish undergo severe muscle tetany which can result in spinal dislocation and muscle haemorrhage (Mazur *et al.*, 1991a). Pulsed white noise, superimposed upon a pulsed carrier, has been used successfully in humans and other terrestrial animals. White noise can be produced by a pulse generator and is then amplified for use by a wide-band power amplifier. Mazur *et al.* (1991b) showed that this was an effective approach in rainbow trout, *Oncorhynchus mykiss*, and coho salmon, *Oncorhynchus kisutch*. Pulsed rectified white noise at 5 to 30 kHz produced anaesthesia with reduced tendency to tetanise the fish, although this was not totally abolished.

In summary

Hypothermia is an effective means of calming animals. While it does not provide real anaesthesia, it does effectively sedate and is suitable for transportation. Electroanaesthesia provides good short-term **apparent** sedation and anaesthesia, although this is not true anaesthesia. While this enables handling and minor procedures, the data available to date suggest that analgesia is good over a quite limited period. The technique has some inherent safety problems, although these can be partly addressed electronically and further enhanced by adopting a working protocol. The tetany induced during some forms of electroanaesthesia can be moderated using pulsed white noise. This technique is clearly attractive in many circumstances but its mechanisms and details of its management are imperfectly understood. There is much scope for further work in this promising area.

References

Anon (1982). Eel exports. *Fish Farmer.* 5 (4): 29.

Barham, W.T. and Schoonbee, H.J. (1990). Induction behaviour of the tilapia *Oreochromis mossambicus* Peters (Pisces: Cichlidae) subjected to electronarcosis by various alternating or rectified currents. *Water South Africa.* 16 (1): 75-78.

Barham, W.T. and Schoonbee H.J. (1991a). A comparison of the effects of alternating current electronarcosis, rectified current electronarcosis and chemical anaesthesia on the blood physiology of the freshwater bream *Oreochromis mossambicus* (Peters). 2. The effect on haematocrit, haemoglobin concentration, red cell count, mean cell volume, mean cell haemoglobin and mean cell haemoglobin concentration. *Comparative Biochemistry and Physiology.* 98A (2): 179-183

Barham, W.T. and Schoonbee, H.J. (1991b). A comparison of the effect of alternating current electronarcosis, rectified current electronarcosis and chemical anaesthesia on the blood physiology of the freshwater bream (tilapia) *Oreochromis mossambicus* (Peters). 3. The effect on the plasma electrolytes Ca^{2+}, Na^{+} and K^{2+} and on the osmotic pressure of the blood. *Comparative Biochemistry and Physiology.* 100A (2): 357-360.

Barham, W.T., Schoonbee, H.J. and Visser, J.G.J. (1987a). The use of electronarcosis as anaesthetic in the cichlid *Oreochromis mossambicus* (Peters). I. General experimental procedures and the role of fish length on the narcotising effects of electric currents. *Onderstepoort Journal of Veterinary Research.* 54: 617-622.

Barham, W.T., Schoonbee, H.J. and Visser, J.G.J. (1987b). The use of electronarcosis as anaesthetic in the cichlid *Oreochromis mossambicus* (Peters). II. The effects of changing physical and electrical parameters on the narcotising ability of alternating current. *Onderstepoort Journal of Veterinary Research.* 55 (4): 205-215.

Barham, W.T., Schoonbee, H.J. and Visser, J.G.J. (1989). Some observations on the narcotising ability of electric currents on the common carp *Cyprinus carpio. Onderstepoort Journal of Veterinary Research.* 56 (3): 215-218

Bell, G.R. (1964). A guide to the properties, characteristics and uses of some general anaesthetics for fish. *Bulletin of the Fisheries Research Board of Canada.* No. 148.

Cowx, I.G. (1990). *Developments in Electric Fishing.* Fishing News Books. Oxford, 358pp.

Cowx, I.G. and Lamarque, P. (1990). *Fishing with Electricity. Applications in Freshwater Fishery Management.* Fishing News Books, Oxford. 248pp.

Darroux, F. (1983). Use of electroanaesthesia in rainbow trout. M.Sc. thesis. University of Stirling. 52pp.

Ellis, J.E. (1975). Electrotaxic and narcotic responses of channel catfish to various electric pulse rates and voltage amplitudes. *Progressive Fish Culturist.* 37 (3): 155-157.

Fish, F.F. and Hanavan, M.G. (1948). A report on the Grand Coulee fish maintenance project (1939-1947). *U.S. Fish and Wildlife Service Special Scientific Report.* No. 55. 63pp.

Gunstrom, G.K. and Bethers, M. (1985). Electrical anaesthesia for handling large salmonids. *Progressive Fish Culturist.* 47 (1): 67-69.

Halsband, E. (1967). Basic principals of electric fishing. In: *Fishing with Electricity - its Applications to Biology and Management* (Edited by R. Vibert). Fishing News Books, Oxford. 276pp.

Holzer, W. (1932). Uber die Stromdichte in Forelleri bei galvanischer Durchrommung in Flussigkecht. *Pflugers Archiv.* 232 (6): 835-841.

Kolz, M.L. (1989). Current and power determinations for electrically anaesthetised fish. *Progressive Fish Culturist.* 51 (3): 168-169

Kynard, B. and Lonsdale, E. (1975). Experimental study of galvonarcosis for rainbow trout (*Salmo gairdneri*) immobilisation. *Journal of the Fisheries Research Board of Canada*. 32: 300-302.

Lamarque, P. (1990). Electrophysiology of fish in electric fields. In: *Fishing with Electricity. Applications in Freshwater Fishery Management* (Edited by I.G. Cowx and P. Lamarque). Fishing News Books, Oxford. 248pp.

Larson, L.L. (1995). A portable restraint cradle for handling large salmonids. *North American Journal of Fisheries Management*. 15 (3): 654-656.

Ludwig, N. (1930). Uber electrotaxis und elektronarkose von Fischen. *Pflugers Archiv*. 244 (2): 268-277.

Madden, J. and Houston, A. (1976). Use of electroanaesthesia with freshwater teleosts; some physiological consequences in the rainbow trout, *Salmo gairdneri*. Richardson. *Journal of Fish Biology*. 9 (6): 457-462.

Mazur, C.F., Boreham, A., McLean, W. and Iwama, G.K. (1991a). Improvements on the use of two alternatives to chemical anaesthesia for fish: Electroanaesthesia and CO_2 anaesthesia. ICES-Council papers. ICES-CM-1991/F:25. 1pp.

Mazur, C.F., Boreham, A., McLean, W. and Iwama, G.K. (1991b). Rectified wide band white noise as an electroanaesthesia waveform for use with rainbow trout (*Oncorhynchus mykiss*). ICES-Council papers. ICES-CM-1991/F:26. 28pp.

Okoye, R.N. (1982). Techniques for transportation of juvenile tilapia. M.Sc. thesis, University of Stirling. 39pp.

Orsi, J.A. and Short, J.W. (1987). Modifications in electrical anaesthesia for salmonids. *Progressive Fish Culturist*. 49 (2): 144-146.

Parker, G.H. (1939). General anaesthesia by cooling. *Proceedings of the Society for Experimental Biology and Medicine*. 42: 186-197.

Randall, D.J. and Hoar, W.S. (1971). Special techniques. In: *Fish Physiology*. Vol. 6. (Edited by W.S. Hoar and D.J. Randall). Academic Press, New York. 559pp.

Robinson, E. (1984). A study of the use of alternating current for electroanaesthesia in *Salmo gairdneri* and *Oreochromis niloticus*. B.Sc. Thesis, University of Stirling. 26pp.

Ross, B. and Ross. L.G. (1984). *Anaesthetic and Sedative Techniques for Fish*. Institute of Aquaculture, University of Stirling. 42pp.

Scheminsky, F. (1934). Uber die Naturder Wechselstrommnarkose bei Fischen. *Arbeiten der Ungarische Biologische Forschungsinstitut*. 6: 209-211.

Schreck, C.B., Whaley, R.H., Bass, M.L., Maglon, O.E. and Solazzi, M. (1970). Physiological responses of rainbow trout (*Salmo gairdneri*) to electroshock. *Journal of the Fisheries Research Board of Canada*. 33 (1): 76-84.

Solomon, D.J. and Hawkins, A.D. (1981). Fish capture and transport. In: *Aquarium Systems* (Edited by A.D. Hawkins). Academic Press, London.

Vibert, R. (1967). *Fishing with Electricity.* Fishing News Books, Oxford. 276pp.

Williamson, R.M. and Roberts, B.L. (1981). Body cooling as a supplement to anaesthesia for fishes. *Journal of the Marine Biological Association of the U.K.* 61 (1): 129-131.

Yoshikawa, H., Ueno, S. and Mitsuda, H. (1989). Short- and long-term cold-anaesthesia in carp. *Bulletin of the Japanese Society of Scientific Fisheries.* 55 (3): 491-498.

~~~~

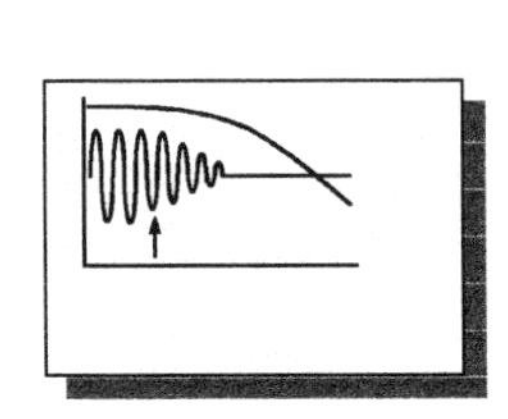
~~~~

Chapter 11

Anaesthesia of Amphibians and Reptiles

Introduction

Anaesthesia of amphibians and reptiles is likely to be practised for research and veterinary work, but only to a very limited extent in aquaculture. There are, however, some circumstances where the aquaculturist may wish to use these techniques. There is little information on anaesthesia of these animals other than for experimental purposes and so the following accounts are, necessarily, brief. The data presented is nevertheless based on practical experiences and the techniques have been proven to be effective. Both phyla consist of animals which are essentially terrestrial, although, with the exception of snakes, lizards and tortoises, much of their lives may be spent in or near water. Both phyla have lungs and so the principal approaches to anaesthesia are by gaseous inhalation or by injectable drugs, although drugs can also be absorbed through the skin in frogs.

Amphibians

General precautions

As amphibians are poikilotherms, care must be taken to ensure that they do not overheat or cool down excessively. The skin of frogs is very prone to desiccation and drying must be prevented. This can be achieved by using a moistened container, or by direct spraying with water. Maintenance of temperature and humidity not only ensures survival but also maximises drug absorption and excretion. Amphibians can be restrained manually, with wet hands, for short periods, although nets may be needed for fully aquatic species and care should be taken to avoid skin damage. During anaesthesia, ventilation of the lungs may cease, but the cutaneous circulation is well adapted in these animals and seems to be sufficient to maintain resting metabolism, especially at low temperatures (Crawshaw, 1992).

Anaesthetic gases and agents in solution are rapidly absorbed and transferred through the skin of amphibians, providing a simple route for anaesthetic administration (Green, 1979). Amphibia can be placed in a moist atmosphere containing inhalational drugs, immersed in a drug solution or wrapped in a cloth moistened with a drug solution and these are the preferred methods for adult frogs. Administration of drugs by spraying onto the abdomen is also a relatively low-stress method for frogs (Fig. 32). Intraperitoneal injection is also effective and intravascular injection can be easily made into the paired dorsal lymph sacs at the base of the last vertebra. The depth of anaesthesia can be judged from the abolition of the withdrawal reflex when the digits are gently pinched, a diminution of the eye retraction reflex and a reduction of postural tone in the muscles of the forelimbs. Frogs should not be allowed to recover in water as they may drown.

Local anaesthesia

Topical anaesthesia is effectively achieved using 2% lignocaine (Lidocaine, Xylocaine) (Johnson, 1992). This can be applied directly by swab or spray (Fig. 32) and is useful for minor procedures.

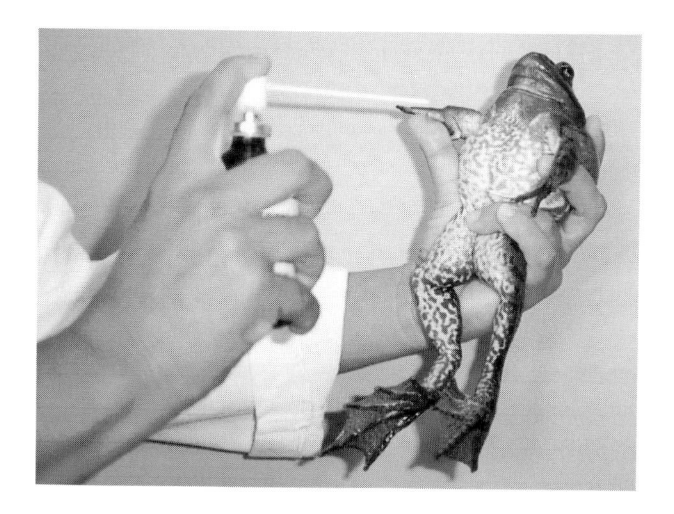

Fig. 32. Delivering Xylocaine as a sedative by metered spray to the abdomen of the bullfrog, *Rana catesbiana.*

Lignocaine applied to and absorbed through the skin by metered spray can also be used as a more general anaesthetic and lengthy deep sedation can be achieved in both *R. pipiens* and *R. catesbiana*. Induction is slow and care must be taken not to overdose because of this.

Gaseous inhalation anaesthesia

Rana pipiens can be anaesthetised using ether (diethyl ether) in a bell jar or similar container (Lumb and Jones, 1973). Respiration is depressed and recovery prolonged.

Methoxyflurane is a useful material which can be administered on a ball of cotton wool placed in a suitable glass or clear plastic container. Johnson (1992) notes that 0.5 to 1.0 cm^3 of 3% methoxyflurane per litre of container induces surgical anaesthesia in about 2 minutes and this is maintained for about 30 minutes. Full recovery is rather prolonged, requiring about 7 hours.

Immersion techniques

This is probably the most effective and easiest technique for use with amphibians, depending upon the absorptive characteristics of the well-vascularised skin.

Immersion in a 1% urethane solution (ethyl carbamate) induces anaesthesia in frogs in 4 to 5 minutes and lasts for some hours after removal from the solution (Westhues and Fritsch, 1965). Although not an immersion technique, urethane can also be spread as a powder onto the back of frogs. It is then absorbed through the skin and anaesthesia commences after 8 minutes, at which point the powder must be washed off the animal. The effect can last for up to 3 hours. Users are cautioned concerning the carcinogenic properties of urethane.

Ethanol (10%) applied to the skin induces anaesthesia in about 10 minutes. After removal from the drug, anaesthesia last for about 20 minutes and recovery requires about 40 minutes (Kaplan and Kaplan, 1961). Overdoses lead easily to fatalities.

Frogs are anaesthetised after 4 to 8 minutes in 0.2% chlorbutanol (Lumb and Jones, 1973). Recovery is slow and takes 3 to 6 hours. Frangioni and Borgioli (1991) used chlorbutanol successfully with the newt *Triturus cristus carnifex*.

Thienpont and Niemegeers (1965) showed that 16 ppm propoxate anaesthetised *Rana esculenta* after 10 minutes.

MS222 is probably the most widely used immersion anaesthetic for amphibians. A 0.5 to 1% solution of MS222 is widely used as a bath after buffering with sodium bicarbonate (Crawshaw, 1992), although much lower doses can also be effective. Adult *Rana pipiens* and *Rana temporaria* and *Xenopus laevis* can be anaesthetised using solutions of about 0.1%, induction requiring about 5 minutes and recovery 40 minutes, although this is prolonged if the maintenance period is extended. Tadpoles can be safely induced using only 0.03% solutions.

Crawshaw (1992) notes that benzocaine is also effective with amphibians and is considered superior by some workers. Induction requires about 5 minutes and recovery about 15 to 30 minutes. Xylocaine works well as a sedative agent in *Rana catesbiana* and *Rana pipiens* (Ross, unpublished observations). Administration as one or two actuations of a 10% Xylocaine metered spray system (Fig. 32) induces a drowsy state within 1 to 2 minutes. Sedation satisfactory for routine handling lasts for about 10 minutes with a progressive recovery.

Injectable drugs

MS222 is the most widely used amphibian injectable drug. It is used as a 0.1 to 0.5% solution at a rate of about 13 $mg.kg^{-1}$ and gives induction in 5 minutes with recovery after 15 to 30 minutes. Letcher (1992) reported that use of intracoelomic MS222, at dosages of 100, 250, and 400 $mg.kg^{-1}$, rapidly induced tranquillisation or anaesthesia in *Rana catesbiana* and *Rana pipiens*, effects being less pronounced or non-existent at 50 $mg.kg^{-1}$. Depth and duration of anaesthesia were dosage related, the greatest depth of anaesthesia in *Rana pipiens* being attained at the 100, 250 and 400 $mg.kg^{-1}$ dosages. This species remained anaesthetised for a significantly greater duration than did *R. catesbiana*. Overall, dosages of between 250 and 400 $mg.kg^{-1}$ reliably induced deep anaesthesia without mortality in bullfrogs.

Barbiturates give a range of responses. Kaplan *et al.* (1962) reported that hexobarbitone injected into *Rana pipiens* at 120 $mg.kg^{-1}$ induced anaesthesia in 20 minutes, lasting for 9 hours, whereas pentobarbitone (Nembutal) at 60 $mg.kg^{-1}$ induced in 18 minutes and lasted for 9.5 hours.

Etorphine at 0.25 mg per animal gave anaesthesia with good analgesia in *Rana pipiens*, lasting for 6 to 12 hours (Wallach and Hoessle, 1970) and 0.5 cm^3 of 5% procaine hydrochloride induced anaesthesia in *Rana catesbiana* in 3 to 5 minutes, persisting for 1 hour (Kisch, 1947).

Dose rates for amphibian anaesthesia are summarised in Table 12.

Table 12. Summary of anaesthetic drugs for use with amphibians. Data drawn from numerous authors.

Mode	Drug	Species or stage	Dose
Inhalation	Ether (diethyl ether)	Frogs	–
	Methylfluorane	Frogs	0.5-1%
Immersion	Benzocaine	Larvae	50 $mg.l^{-1}$
		Frogs and salamanders	200-300 $mg.l^{-1}$
	Chlorbutanol	Frogs	0.2%
	Ethanol	Frogs	10%
	MS222: bath	Tadpoles and newts	200-500 $mg.l^{-1}$
		Frogs and salamanders	500-2000 $mg.l^{-1}$
		Toad	1000-3000 $mg.l^{-1}$
	Propoxate	Frogs	16ppm
	Urethane	Frogs	1%
Injection IM or IP	Etorphine (Immobilon)	*R. pipiens*	0.25mg/frog
	Hexobarbitone	*R. pipiens*	120 $mg.kg^{-1}$
	MS222	*R. pipiens* *R. catesbiana*	100-400 $mg.kg^{-1}$ 250-400 $mg.kg^{-1}$
	Pentobarbitone (Nembutal)	*R. pipiens*	60 $mg.kg^{-1}$
	Procaine (5%)	*R. catesbiana*	0.5 cm^3 /frog

Hypothermia

Amphibians are poikilotherms and can be immobilised by lowering their body temperature. This may be useful for handling but it should be remembered that this is not real anaesthesia and that no analgesia will be provided. No guidelines are available on safe temperature reductions in amphibians and so the technique should be used with care to avoid mortality.

Reptiles

The principal management constraint with reptiles is, again, temperature, and steps must be taken to avoid overheating. The problems of skin desiccation do not apply as the integument is heavily keratinised and this presents another problem for the anaesthetist. Reptiles are, strictly speaking, terrestrial animals and those reptiles of major interest to aquatic biologists and aquaculturists are confined to turtles and crocodiles. The snakes and lizards are omitted from this discussion.

Crocodilians

General precautions

Crocodiles and alligators can present some considerable danger to operators and any attempt at anaesthesia must be carefully considered in advance to avoid wounding by bites or mechanical abrasion with the scaly skin of crocodilians. It may be necessary to wear strong protective clothing and gloves should be used. Crocodiles and alligators are problematic and the jaws, body and the tail must all be restrained. In smaller animals (approximately up to 50 cm) this can be easily achieved by a skilled assistant using gloved hands on the snout and tail. Above this size direct handling becomes progressively more difficult and may require several assistants. In these cases jaw restraint must be managed remotely by using a lasso and body restraint by assistants. With large crocodiles, use of a dart gun with etorphine may be preferable, avoiding direct contact with the unanaesthetised animal altogether.

Crocodiles and alligators metabolise drugs quite slowly and hence recovery is slow.

Local anaesthesia

As in amphibians, topical anaesthesia can be achieved using 2% lignocaine applied directly by swab or spray (Johnson, 1992). This is, however, more effective in snakes and lizards than turtles and crocodiles due to the nature of the integument.

Oral administration

Pentobarbitone induced deep anaesthesia when fed to small alligators at 45 $mg.kg^{-1}$ (Pleuger, 1950).

Injectable drugs

Brisbin (1966) investigated the use of a number of agents with alligators. Pentobarbitone administered intramuscularly to 2 to 5 kg *Alligator mississippiensis* at 8 $mg.kg^{-1}$ gave a recovery period of 2 to 3 hours. MS222 at 80 to 100 $mg.kg^{-1}$

immobilised after 10 minutes, with a 9 to 10 hour recovery period. Phencyclidine is effective within 60 minutes in alligators at 12 to 24 $mg.kg^{-1}$, with a 6 to 7 hour recovery period. Wallach *et al.* (1967) showed that intramuscular etorphine sedates crocodiles at a rate of 0.3 $mg.kg^{-1}$.

Ketamine is the major drug in current use with crocodilians. It is effective at intramuscular weight-related doses between 12 and 80 $mg.kg^{-1}$ in a range of animals. Induction takes 30 minutes or more, the effect may last for 4 to 7 hours and full recovery can be lengthy.

Johnson (1992) notes that telazol (tiletamine hydrochloride and zolazepam hydrochloride) may be preferable to ketamine as recovery is shorter. It is also effective at a lower dose (10 to 30 $mg.kg^{-1}$) but the working solution does not keep well.

Morgan-Davies (1980) described the use of gallamine triethiodide with *Crocodilus niloticus*. He obtained immobilisation within 15 minutes and considered that it was preferable to etorphine or phencyclidine because of its availability in the region (West Africa). Suxamethonium chloride (scoline) gave complete muscle relaxation in alligators at 3 to 9 $mg.kg^{-1}$ within 4 minutes, subsequent recovery requiring 7 to 9 hours (Brisbin, 1966). Similarly, Messel and Stephens (1980), using suxamethonium, obtained good immobilisation of *Crocodilus porosus* and *C. johnsoni* within 7-12 minutes. They noted that *C. porosus* required ten times the weight-related drug dose to achieve the same effect and that the drug was ineffective if the head was kept below water. This phenomenon may be related to the vascular shunting which occurs in the submerged animal. Although both gallamine and scoline are muscle relaxants which can facilitate handling, they are not full anaesthetics.

Turtles and tortoises

General precautions

Turtles and tortoises have a low metabolic rate and anaesthesia and recovery can be expected to be quite slow. They can be handled for anaesthesia if care is taken to avoid bites and this can be simply achieved by wearing appropriate clothing and gloves. Particular care is needed in the case of the soft-shelled turtles, which can move very quickly. In some cases tongs or ropes may be needed for restraint and in aggressive freshwater species MS222 can be added to the water as a tranquilliser.

Oral or rectal anaesthesia

Drugs can be given by mouth or by rectal instillation, although turtles given urethane by mouth at 2.73 $g.kg^{-1}$ were not anaesthetised reliably. Oral administration of pentobarbitone to red-eared turtles, *Pseudemys scripta*, was also considered unpredictable (Vos-Maas and Zwart, 1976). Rectally instilled tribromoethanol, given with amylene hydrate (tertiary amyl alcohol), induced tortoises in 1 to 2 hours and anaesthesia was maintained for over an hour (Hunt, 1964).

Gaseous inhalation anaesthesia

Anaesthesia can be obtained in turtles and tortoises using some fairly simple compounds. Lumb and Jones (1973) administered ether to turtles using a face mask. Induction required 30 to 40 minutes and recovery up to 10 hours. More recently, Bello and Bello-Klein (1991) administered ether through a plastic cannula placed in the trachea to anaesthetise the turtle *Chrysemys dorbigni* for up to 90 minutes, with recovery requiring 90 minutes and with no recorded deaths. Methoxyflurane has also been used widely in reptilian work. Induction with a 3% gas:air mixture takes about 8 to 25 minutes and recovery is variable. A 1.5% mixture can be used for maintenance of anaesthesia. Halothane can also be used for induction and maintenance at the same concentrations and gives induction in 1 to 3 minutes. Full recovery can be very prolonged, however, and may require up to 24 hours. Johnson (1992) notes that isofluorane is effective at similar doses but that recovery requires only about 3 hours, giving this compound some advantages.

Injection anaesthesia

Turtles have a thick skin and sharp needles must be used at all times. Intramuscular injections can be made into the muscles of the hind limb, while intravascular injection is into the heart or the ventral abdominal vein, both of which are technically difficult and not advised. Intraperitoneal routes are via the soft skin at the base of the limbs.

Pentobarbitone given to turtles by intracardiac injection at 10 $mg.kg^{-1}$ induced anaesthesia in 15 minutes, surgical levels being maintained for 3 hours (Young and

Kaplan, 1960). Intraperitoneal pentobarbitone at 18 mg.kg^{-1} induced sleep in tortoises after 18 minutes and surgical anaesthesia within 80 minutes with prolonged recovery (Hunt, 1964). Ketamine given intramuscularly at 60 mg.kg^{-1} gives light anaesthesia after 30 minutes which lasts for about 60 minutes. Although full recovery may take up to 24 hours, this is probably the method of choice.

Dose rates for reptilian anaesthesia are summarised in Table 13.

Hypothermia

All reptiles are poikilotherms and can be immobilised by lowering their body temperature. Low temperature also gives some haemostasis. Few guidelines are available on safe temperature reductions in reptiles. However, animals can be placed in a refrigerator or in crushed ice. Recovery is usually rapid and deaths are rare. This technique is useful for handling but it should be remembered that this is not real anaesthesia and that no analgesia will be provided. If surgery is contemplated, therefore, a local anaesthetic should be used at the site.

Table 13. Summary of anaesthetic drugs for use with "aquatic" reptiles. Data drawn from numerous authors. IM = intramuscular, IP = intraperitoneal.

Mode	Drug/mode	Species	Dose (mg.kg^{-1})	Induction (mins)	Recovery (hours)
Oral	Pentobarbitone	Crocodilians	45	-	-
	Urethane	Turtles	2300	-	-
Injection	Etorphine	Crocodilians	0.3	-	By antidote
	Gallamine triethiodide: IM	Crocodilians		7-12	
	Ketamine: IM	Crocodilians, turtle/tortoise	12-25 40-80	30-60 30	4-7 >24
	MS222: IM	Crocodilians	80-100	10	9-10
	Pentobarbitone: IM	Crocodilians, turtles	8-10 20-50	15-18	3
	Pentobarbitone: IP	Turtles	18	18	Long
	Phencyclidine: IP	Crocodilians	12-24	60	6-7
	Succinylcholine chloride: IM	Crocodilians, turtle/tortoise	0.5-2 0.1-1	20-30 20-30	
	Tiletamine-zolazepam: IM	Crocodilians, turtle/tortoise	15 10-25	>20 5-20	
Inhalation	Ether (diethyl ether)	Turtles	to effect	-	1.5
	Halothane	General	3%	1-30	7-24
	Isoflurane	General	3%	1-15	3
	Methoxyflurane	General	3%	8-25	Variable

In summary

The methods of choice for amphibians will depend upon circumstances, but the wide range of options means that there will be little difficulty in adopting a method. Crocodilians are best anaesthetised by injection and the handling problems inherent in this call for caution. Initial handling for injection calls for some practice in the use of nooses, ropes, etc. Where larger animals are the subject of investigations there may be little option other than to use a dart gun. Turtles can also be difficult to handle and can inflict severe wounds by biting. The ability to use inhalation as an alternative to injection may be attractive for this reason but it should be remembered that the low metabolic rates will give very slow response times.

References

Bello, A.A. and Bello-Klein, A. (1991). A technique to anaesthetise turtles with ether. *Physiology and Behaviour.* 50 (4): 847-848.

Brisbin, I.J.L. (1966). Reactions of the American alligator to several immobilising drugs. *Copeia.* 129-130.

Crawshaw, G.J. (1992). Medicine and diseases of amphibians. In: *The Care and Use of Amphibians, Reptiles and Fish in Research. Proceedings of a SCAW Conference* (Edited by D.O. Schaeffer, K.M. Kleinow and L. Krulisch). Scientists Center for Animal Welfare, Bethesda, MA. 196pp.

Frangioni, G. and Borgioli, G. (1991). Effect of spleen congestion and decongestion on newt blood. *Journal of Zoology.* 223 (1): 15-25.

Green, C.J. (1979). *Animal Anaesthesia.* Laboratory Animal Handbooks. No. 8. Laboratory Animal Ltd, London. 300pp.

Hunt, T.J. (1964). Anaesthesia of the tortoise. In: *Small Animal Anaesthesia* (Edited by O. Graham-Jones). Pergamon, Oxford. pp. 71-76.

Johnson, H.J. (1992). Anaesthesia, analgesia and euthanasia in reptiles and amphibians. In: *The Care and Use of Amphibians, Reptiles and Fish in Research. Proceedings of a SCAW Conference.* (Edited by D.O. Schaeffer, K.M. Kleinow and L. Krulisch). Scientists Center for Animal Welfare, Bethesda, MA. 196pp.

Kaplan, H.M. and Kaplan, M. (1961). Anaesthesia in frogs with ethyl alcohol. *Proceedings of the Animal Care Panel.* 11: 31-36.

Kaplan, H.M., Brewer, N.R. and Kaplan, M. (1962). Comparative value of some barbiturates for anaesthesia in the frog. *Proceedings of the Animal Care Panel.* 12: 141-148.

Kisch, B. (1947). A method to immobilise fish for cardiac and other experiments with procaine. *Biological Bulletin of the Marine Biological Laboratory. Woods Hole.* 93: 208.

Letcher, J. (1992). Intracelomic use of tricaine methanesulfonate for anesthesia of bullfrogs (*Rana catesbeiana*) and leopard frogs (*Rana pipiens*). *Zoological Biology.* 11 (4): 243-251.

Lumb, W.V. and Jones, E.W. (1973). *Veterinary Anaesthesia.* Lea and Febiger. Philadelphia.

Messel, H. and Stephens, D.R. (1980). Drug immobilization of crocodiles. *Journal of Wildlife Management.* 44 (1): 295-296.

Morgan-Davies, A.M. (1980). Immobilization of the Nile crocodile (*Crocodilus niloticus*) with gallamine triethiodide. *Journal of Zoology and Animal Medicine.* 11 (3): 85-87.

Thienpont, D. and Niemegeers, C.J.E. (1965). Propoxate (R7464): A new potent anaesthetic agent in cold-blooded vertebrates. *Nature.* 205: 1018-1019.

Vos-Maas, M.G. and Zwart, P. (1976). A technique for intravenous injection in the red-eared turtle (*Pseudemys scripta elegans*). *Laboratory Animals.* 10: 399-401.

Wallach, J.D. and Hoessle, C. (1970). M99 as an immobilising agent in poikilotherms. *Veterinary Medicine - Small Animal Clinician.* 65: 161-162.

Wallach, J.D., Frueh, R. and Lentz, M. (1967). The use of M99 as an immobilising analgesic agent in captive wild animals. *Journal of the American Veterinary Medical Association.* 151: 870-876.

Westhues, M. and Fritsch, R. (1965). *Animal Anaesthesia.* Oliver and Boyd, Edinburgh.

Young, R. and Kaplan, M. (1960). Anaesthesia of turtles with chlopromazine and sodium pentobarbital. *Proceedings of the Animal Care Panel.* 10: 57-61.

~~~~

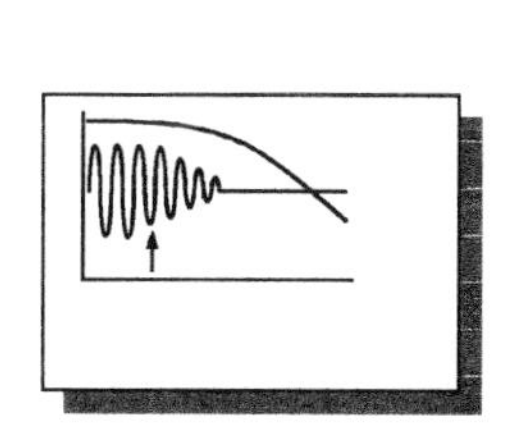
~~~~

Chapter 12
Transportation and Anaesthesia

Introduction

Sedation can be beneficial in bulk transportation of stocks, especially where long distances are to be covered, but there may also be advantages in sedating animals for short journeys. There are four main reasons for transporting aquatic animals:

- The movement of fry or juveniles for stocking into farms or restocking into the natural environment.
- The movement of broodstock for management purposes.
- The movement of live stock to markets.
- The movement of animals for research purposes.

For restocking and aquaculture, the density of transportation may be very high, whereas broodstock and animals for research will usually be transported at a much lower density. Movement of live fish to market is an increasingly important aspect of aquaculture in many parts of the world and, apart from the technical problems of transportation, use of drugs raises a number of important questions where food organisms are being considered.

Some useful reviews of transportation are available (Johnson, 1979; Solomon and Hawkins, 1981; Berka, 1986; Jhingran and Pullin, 1985) which describe many detailed aspects of the technology and approach to live transportation. This chapter discusses the necessary background to the problems of live transportation and focuses on the potential use of anaesthetic and sedative agents in the transportation process.

The physical and biological problems in transportation

The major issues of concern to those wishing to transport aquatic animals are management of handling stress and mechanical shock, potential heat stress and deterioration of water chemistry. The metabolism of animals will affect water chemistry in a closed environment and the deteriorating water quality will, in turn, affect the physiology of the transported animals. This feedback process can lead to rapid degradation of the transport environment. It will readily be appreciated that control of these factors involves some knowledge of the environmental requirements and physiological responses of the animals, enabling management of the transport environment and maintenance of water quality.

Handling

This is an inevitable precursor to any transportation exercise and may cause mechanical abrasion, induce some degree of stress and result in exhausted animals. Visible damage is frequently seen as scale loss, but it should be remembered that the epidermis will have been damaged well before scales are shed. Damage to the delicate epidermal layers and mucous covering will allow invasion by pathogens and disrupt osmoregulation. It should be obvious that any handling should be as careful as possible, and when dealing with broodstock or small numbers of animals this will usually be relatively easy to arrange. The practicalities and pressures of dealing with very large numbers of animals tend to mitigate against careful handling, however, this must be uppermost in the minds of all involved in transportation operations. Alarm pheromones released in crowded conditions may exacerbate the stress level (Malyukina *et al.,* 1982).

Good-quality, preferably knotless nets, containers and other hardware must be used both at the capture and packing stages and must be complemented by the extra time and patience required to ensure safe handling and minimise mortalities. A further source of mechanical abrasion, often overlooked, is from the animals themselves. This is particularly so where they are netted in large numbers or carried at high density.

Mechanical stress

During the transportation process itself, there will certainly be some level of vibration, sudden shocks and perhaps a greatly elevated noise level, particularly in vehicles. All of these factors will directly cause a stress reaction in most animals and are known to induce a cortisol reaction in fish. In view of the fact that vehicles must be used and the condition of roads is variable, there is often little that can be done to ameliorate these influences, other than be aware of the potential stress they may cause.

Interestingly, Laird and Wilson (1979) showed survival benefits from transporting salmonid eggs in a solution of 1% methylcellulose, its viscosity presumably providing some buffering from mechanical shock.

Temperature

Temperature affects all aspects of metabolism and at higher temperatures metabolic rate increases. In general, the transportation water temperature should not be allowed to rise above the starting temperature and should be reduced if possible, within certain limits. This is not easy to achieve in practice, except where specialist equipment is being used, for example using insulated tanks on purpose-built fish transportation trucks. There are, however, numerous examples of relatively simple cooling and insulation techniques in daily use around the world. The cooling effect of evaporation from earthenware pots and wet sacking is a frequently used tool in the Far East for small-scale work. Insulated containers, particularly injection-moulded polystyrene (styrofoam) boxes, are now very widespread and are extremely effective.

There are many ways of using ice, either by direct addition to the medium or by packing fish in plastic bags into crushed ice, the latter being effective for animals requiring saline environments. Chipped or crushed ice is best, and flaked ice probably melts too quickly. Ice blocks can be stored in a deep freeze ready for use. The permanent, recyclable, freezable plastic bricks which are widely available for food use have found many applications and, particularly during transportation by air, can provide a non-chemical, safe, reduced temperature environment for many hours. Cooling can also be achieved by the addition of dry ice (solid carbon dioxide at -73°C) in thermal contact with the water but chemically isolated from it (Solomon and Hawkins, 1981).

On the other hand, thermal shock caused by over-cooling must be avoided and few species will tolerate an instantaneous temperature reduction of more than 10°C. As described earlier, tolerable reductions which guarantee no attendant mortalities may be as little as 5 to 6°C.

Dissolved oxygen

Oxygen is the first **limiting** component of the aquatic environment, other than water itself, and ensuring the oxygen supply of transported animals is highly important. Many factors affect the respiratory rate of animals, including stress, temperature, salinity, body weight and feeding level. Oxygen consumption will be high during the earlier part of a journey due to initial stress and this period particularly must be carefully managed. Temperatures should certainly not be allowed to rise and should be reduced if it is practical to do so. Salinity should be maintained or brought towards the iso-osmotic point where possible to minimise the osmoregulatory work load. The minimum tolerated oxygen level varies with species. Tilapias, for example, are relatively hardy and will tolerate transportation D.O. down to 3 $mg.l^{-1}$, whereas salmonids respond badly below about 5 $mg.l^{-1}$, or higher. Feeding increases metabolism due to the requirements for biochemical processing, an effect known as specific dynamic action, or SDA. This can increase metabolic rate and oxygen consumption by more than 50% but fortunately can be minimised during transportation by the simple expedient of starving the animals prior to travel. The starvation period depends upon the gut evacuation rate, which is species and temperature dependent. Typical starvation periods vary from 12 to 24 hours but can be longer in cold-water species with long gut evacuation times.

Salinity

The body fluids of most animals have a concentration between that of fresh water and sea water, usually approximately one-third the strength of sea water. This internal environment is regulated to a greater or lesser degree in different animals, but is very important for proper functioning, health and survival. Salinity is the second most important environmental variable influencing the rate of metabolism. In many cases the salinity of the transportation medium will be identical to the source of the animals. However, it should be borne in mind that salinities of about 15‰ are more iso-osmotic and will reduce osmoregulatory work. Thus, in those animals which can tolerate it, a reduction or increase in salinity towards this iso-osmotic point can notably reduce metabolic rate and hence oxygen consumption and ammonia excretion. Simple dilution of the environment, or addition of sodium chloride or, preferably, sea salts can bring about the required adjustment. Extreme care must be taken in attempting to use this approach with stenohaline animals which do not normally survive such changes well; most carps are good examples of species which cannot benefit as they will tolerate only a few parts per thousand salinity.

Carbon dioxide

Carbon dioxide can accumulate rapidly in the transportation medium, especially at high stocking densities. This may cause some stress in the animals and above about 40 $mg.l^{-1}$ it may be fatal. However, it may also have a mild sedative effect at low levels and this can be beneficial. Carbon dioxide levels will be minimised if air bubbling is used for oxygenation, and the container is adequately vented to the outside. At higher densities, however, little can be done to prevent its accumulation.

Ammonia

This constitutes 70%, or more, of the nitrogenous wastes from aquatic organisms. It is excreted directly, and continuously, into the transporting medium, principally via the gills. There is very little urinary excretion of ammonia and hence no periodic elimination, as found in terrestrial animals. Ammonia is very soluble in water and in the unionised state, as NH_3, it is very toxic, whereas in the ionised state, as NH_4, it is not toxic. The state of the ammonium ion depends upon the pH of the environment and to a lesser extent the temperature. Higher pH and lower temperatures cause greater ionisation, and hence higher toxicity. Low oxygen levels increase toxicity, although this can be ameliorated by high carbon dioxide concentrations which produce a concomitant reduction in pH. The rate of ammonia excretion is determined by feeding level, dietary protein content and temperature. Consequently, ammonia excretion can be reduced markedly by cooling and withdrawing food from animals to be transported. High pH waters can be buffered down to reduce toxicity.

Suspended solids

Suspended solids will irritate the gills, obscure the water and cause stress. Again, the source of suspended solids is faeces and prior withdrawal of food is the cure. Simple filtration may also be possible in more complex transportation situations.

Basic techniques for transportation

The important issues facing aquaculturists and biologists wishing to transport animals are:

- conditioning
- containers
- control
- sedation

Conditioning

The initial phase is capture or extraction from the holding facility, involving netting and handling, which have already been discussed. This is often followed by some form of conditioning, which usually consists of a period of starvation in holding facilities with good water exchange to ensure that all faeces are voided and that no solids or metabolites are transferred with the animals. Sampson and Macintosh (1986) note the importance of this conditioning treatment. Jhingran and Pullin (1989) recommended applying additional stress to ensure defaecation in carp fry. While there is a certain logic here, this procedure cannot be recommended as a general practice for most aquatic animals and an adequate conditioning period will ensure normal gut evacuation without recourse to such practices. During this period the animals will also substantially overcome the stress of temporarily being held at higher densities. Finally, the animals are placed in the transportation containers.

Containers

There is an enormous range of transportation containers, from simple to complex. As mentioned earlier, in addition to transporting the animals in the minimum amount of water, these containers may be designed to facilitate maintenance of temperature, exchange of gases and perhaps exchange of water.

Many varieties of open, small-scale, simple containers are used in rural settings where small numbers of animals are to be carried, or even large numbers of very young larvae. The containers can be made of plastic, wood or metal. As mentioned earlier, earthenware pots or wetted sacking placed over simple containers can provide significant evaporative cooling. These allow rapid water exchange if needed and the water surface is often continuously agitated by hand to introduce oxygen; tedious but effective. Long journeys on foot or bicycle are often made using only this technology and empirical procedures to ensure maximum survival are well know to the practitioners and journeys are often carefully planned.

Although sealed containers of plastic, metal or glass have also been used, the most widely used small-scale closed container is now the plastic bag (Fig. 33).

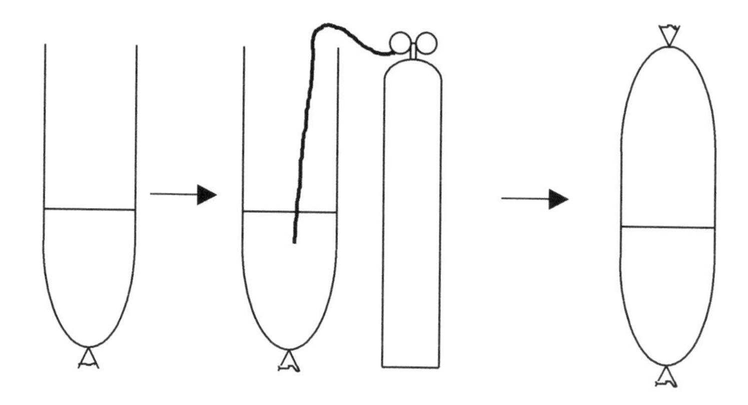

Fig. 33. The plastic bag technique for transportation of fish and crustaceans. Strong bags may be purchased for use, but plastic tube bought as a roll is frequently the method of choice.

In this system, animals are placed in a volume of water filling about 20% of the bag and the remainder of the bag is filled with pure oxygen, which is bubbled through the water during the filling process The bag is then tightly sealed for the journey, but often in such a way that more oxygen could be added if required. In this way a large quantity of oxygen is made available to the animals and extended journeys of 24 hours, or more, can be made. These bags are often used alone or, more frequently, placed in a rigid container to hold them upright. The bag wall provides a smooth, semi-soft surface which reduces impact abrasion, although care must be taken that small animals are not trapped in pocketed folds of the bag, especially at the corners. Plastic bags in polystyrene boxes afford excellent thermal protection and intercontinental air transportation of very large numbers of small fish for the aquarium trade takes place in this way daily. Fish fry and young shrimp for aquaculture also frequently travel in this way. Larger fish or crustaceans can also be carried in plastic bags, either in small numbers or individually.

Large-scale, open containers for transportation of fish fry or adults are often, in fact, semi-closed and will have a vented lid which effectively isolates the transportation medium from the air. These rigid containers may be up to 4 m^3 in volume. The containers themselves are usually well-insulated and are equipped with independent compressed air supplies, or oxygen cylinders, or even both. As large numbers of high-value animals are involved, monitoring systems are often quite advanced, with continuous recording of temperature and dissolved oxygen at least. Although expensive in capital terms, such vehicle-mounted containers are extremely successful, can transport animals over huge distances involving periods of up to 2 days, or more, and are used in the intensive aquaculture industry in many parts of the world.

Controlling the environment

During transportation, use may be made of any number of a range of additives. The water itself may be exchanged, or partly exchanged, at intervals. The addition of oxygen or compressed air has already been discussed and use has been made of hydrogen peroxide as an additional source of oxygen for small-scale work (Marathe *et al.*, 1975; Innes-Taylor and Ross, 1988). Ice or dry ice may be used for cooling. These practices will ensure the basic clean, well-oxygenated, cool environment.

Salts may be added in the form of sodium chloride or calcium chloride to reduce osmotic work. Typical doses range from 1 to 5 $g.l^{-1}$ NaCl (Chittenden, 1971; Wedermeyer, 1972) or 50-75 $mg.l^{-1}$ $CaCl_2$ for freshwater fish. Weirich *et al.* (1992) showed that elevated calcium levels of 80 $mg.l^{-1}$ in the transportation medium boosted survival of *Morone chrysops* x *M. saxatailis* hybrids. Attempts to minimise the level of dissolved ammonia have been made using clinoptilolite or related ion-absorptive resins or activated charcoal, all of which will absorb large quantities of ammonia. One of the difficulties in using these materials is ensuring that they are exposed to the water without being dispersed in it. Many varieties of bag and cartridge systems have been used with some success and water can be simply passed through such containers using air-lift principles. Frequently, antibiotics or disinfectants are used in transportation. These range from methylene blue, malachite green and potassium permanganate, to acriflavin, kanamycin, gentamycin, chloramphenicol, neomycin and furazolidone. Recommended prophylactic doses of these materials vary.

In recent years there has been some interest in using polyvinyl pyrolidone, PVP, for transporting fish. This is a long-chain polymer of varying molecular weight. The belief is that this material coats the skin of fish and helps protect against abrasion. It is also thought to form a coating over any damaged epithelium, so promoting healing. PVP has been promoted as a commercial product in a formulation which also contains extracts of alöe vera. This is also a well-known material, in the traditional or "herbal" medicine category, which has some known anti-microbial properties, as well as being a promoter of wound healing. The scientific evidence for its effectiveness is not yet strongly documented, although the preparation is widely in use for transportaion of aquarium fish and is being evaluated by aquaculturists, especially for deliveries from hatcheries.

Sedation

Apart from minimising handling time, direct physical injury and stress effects during transportation, sedation is useful in reducing metabolic rate and consequently oxygen consumption, and in reducing excretion of metabolic products into the water. The simplest sedation is through cooling; more care is needed when using drugs. A combination of some cooling and low-dose drugs can also be a useful approach. Sedation is by no means an essential ingredient of transportation and the decision to use this approach will depend upon the individual circumstances. The use of drugs, at however low a dose rate, with animals destined for markets must be very carefully considered, along with any necessary customer protection regulations.

Sedation of molluscs

Hovgaard (1985) described a method used in New Zealand for maintaining the keeping quality of the blue mussel, *Perna canalicula,* during transportation. The mussels are cooled by, but are not in direct contact with, ice and the barrier between the ice and mussels allows the melted water to pass through. This method can maintain the mussels in good condition for about 10-12 days.

Panggabean *et al.* (1989) evaluated storage conditions for *Crassostrea gigas* and noted that larvae stored at 5°C had greater survival than larvae held at room temperature.

Solis and Heslinga (1989) showed that survival of seed of *Tridacna derasa* in pure oxygen was higher than that in the ambient atmosphere over exposure times ranging from 16 to 48 hours. They considered that the low cost of the treatment justifies its routine use for commercial seed shipments.

There are no records of the use of sedative drugs in transportation of molluscs. Bivalve molluscs have lengthy survival times in air if kept cool.

Sedation of crustaceans

In many cases, large crustaceans can be transported in air, without the use of anaesthetic drugs, and this is assisted if the animals are kept cool, often in damp seaweed or a similar medium. There are numerous reports of successful transportation using variations of these methods (Meereen, 1991; Meeren and Uglem, 1993) and Whiteley and Taylor (1992) identified the conditioning which affects the success of such operations.

Paterson (1993) describes successful live shipment of *Penaeus monodon* without water following cold "anaesthesia" at 12°C.

Singh *et al.* (1982) showed that chloral hydrate was an effective sedative for the transport of *Penaeus monodon*. A dose of 400 $mg.l^{-1}$ was found to be the most effective dose for successful transportation of 375 $animals.l^{-1}$ for a period of up to 28 hours. Smith and Ribelin (1984) showed that 36 day-old post-larval *P. vannamei* can be packed and shipped at a density of at least 190 $animals.l^{-1}$ for 18 hours at 18-20°C with negligible mortality. Adult *M. rosenbergi* packed unrestricted resulted in survival rates substantially higher than those obtained from prawns immobilised by wrapping in mesh (Smith and Wannamaker, 1983).

Sedation of fish

Yoshikawa *et al.* (1989) examined techniques for cold anaesthesia of *Cyprinus carpio* during transportation and Nakamura (1992) showed that slow cooling of *Cyprinus carpio* prior to cold air transportation at 7°C considerably extended tolerance and increased survival. The sedative effect of carbon dioxide was described earlier. It is an effective agent during transportation but Itazawa and Takeda (1982) and Takeda and Itazawa (1983) showed that, for 100% survival, oxygen levels need to be elevated at the same time as carbon dioxide. This tends to make this approach impractical. Yokoyama *et al.* (1989), who used combined techniques, found that carp could be "anaesthetised" after about 10 minutes at 4°C and could then be transported safely at 14°C in 80 mmHg carbon dioxide.

A number of workers have reviewed tranquillising drugs for transportation, including McFarland (1960), Bell (1964), Durve (1975) and Rothbard (1988). In most cases, only a few agents are recommended and these are summarised in Table 14.

Teo and Chen (1993) noted the importance of using anaesthetics in transportation of aquarium fish by air. The same authors showed suppression of oxygen consumption rates of *Poecilia reticulata* using 2-phenoxyethanol in simulated transportation trials. Based on rate of excretion of metabolic wastes over 48 hours and cost, Guo *et al.* (1993) rated 2-phenoxyethanol best for transportation of aquarium fish, followed by quinaldine, metomidate and MS222.

Benzocaine hydrochloride at 25 $mg.l^{-1}$ was shown to be effective in transporting fish (Ferreira *et al.*, 1984). Okoye (1984) extended this approach and showed that benzocaine anaesthesia combined with moderate cooling produced substantial calming with 100% survival in *Oreochromis niloticus* juveniles. The highest sustainable dose was 32 $mg.l^{-1}$ at 24°C, but combinations of 20 to 25 $mg.l^{-1}$ and 18°C gave best results. Ross and McKinney (unpublished data) showed that the dose of benzocaine which could be used in transportation of rainbow trout was limited by the need to maintain equilibrium and spacing of the animals. If sedated too deeply, fish sink to the bottom of the container, where they may be in high densities locally, resulting in massive and rapid local environmental degradation. Benzocaine at 5 $mg.l^{-1}$ reduced activity for up to 2 hours, while at 7 $mg.l^{-1}$ activity was reduced for 3 hours, although there was some brief initial hyperactivity at this dose. Doses in excess of 10 $mg.l^{-1}$ all eventually resulted in loss of equilibrium and are hence unsuitable for use during long transportation. There was no improvement in water quality over non-sedated fish over 6 hours.

Table 14. Summary of dose rates of drugs used in transportation of fish.

Drug	**Species**	**Dose rate**	**Author**
2-Phenoxyethanol	Salmonids	0.5 ppm 0.25 ppm	Bell (1964) Barton and Helfrich (1981)
	Indian carps	0.3 $cm^3.l^{-1}$	Jhingran and Pullin (1985)
	Aquarium fish	0.2 $cm^3.l^{-1}$	Guo *et al.* (1993)
Amobarbital sodium	Indian carps	85 $mg.kg^{-1}$	Jhingran and Pullin (1985)
Amytal sodium	Indian carps	50-175 $mg.l^{-1}$	Jhingran and Pullin (1985)
Barbital sodium	Indian carps	50 $mg.kg^{-1}$	Jhingran and Pullin (1985)
Benzocaine	*Oreochromis niloticus*	25 $mg.l^{-1}$	Ferreira *et al.* (1984)
	Oreochromis niloticus	?	Okoye (1982)
Carbon dioxide	*Cyprinus carpio*	95-115 mmHg *pO_2 400+ mmHg	Takeda and Itazawa (1983)
Chloral hydrate	Fish sp.	250 $mg.l^{-1}$	Solomon and Hawkins (1981)
	Indian carps	0.6 $g.l^{-1}$	Jhingran and Pullin (1985)
Etomidate	*Puntius gonionotus*	0.15-0.2 ppm	Gopinath Nagaraj (1990)
Metomidate	*Alosa sapidissima*	0.5-1.0 $mg.l^{-1}$	Ross *et al.* (1993)
	Aquarium fish	1 $mg.l^{-1}$	Guo *et al.* (1993)
MS222	Fish sp.	10-30 $mg.l^{-1}$	Solomon and Hawkins (1981)
	Mugil trichodon	20-30 $mg.l^{-1}$	Alvarez and Garcia (1982)
	Indian carps	10-25 $mg.l^{-1}$	Jhingran and Pullin (1985)
	Cyprinus carpio	20 $mg.l^{-1}$	Horvath *et al.* (1984)
	Hypophthalmichthys molitrix	10 $mg.l^{-1}$	Horvath *et al.* (1984)
	Aquarium fish	30 $mg.l^{-1}$	Guo *et al.* (1993)
	Tinca tinca	5-50 $mg.l^{-1}$	Jirasek *et al.* (1978)
Propoxate	*Tinca tinca*	0.25-0 5 ppm	Jirasek *et al.* (1978)
Quinaldine	Various spp	15-30 ppm	Woynarovich and Horvath (1980)
	Indian carps	25 ppm	Jhingran and Pullin (1985)
	Tilapias	?	Sado (1985)
	Aquarium fish	10 ppm	Guo *et al.*(1993)
Quinaldine + MS222	Fish sp.	5 $mg.l^{-1}$ + 10-30 $mg.l^{-1}$	Solomon and Hawkins (1981)
Tertiary amyl alcohol	Fish sp.	0.3-0.8 $g.l^{-1}$	Solomon and Hawkins (1981)
	Mugil trichodon	0.25-0.5 $cm^3.l^{-1}$	Alvarez and Garcia (1982)
	Indian carps	0.45 $cm^3.l^{-1}$	Jhingran and Pullin (1985)
Tertiary butyl alcohol	*Mugil trichodon*	3.0-3.5 $cm^3.l^{-1}$	Alvarez and Garcia (1982)
Urethane	Indian carps	100 $mg.l^{-1}$	Jhingran and Pullin (1985)

Sado (1985) used quinaldine in transportation of tilapias and a general dose range was 5 to 30 $mg.l^{-1}$.

Kacem *et al.* (1988) experimented with a beta-blocker, carazolol, in transportation of *Dicentrarchus labrax*. They found that red cell counts and haemoglobin levels were higher in fish after a 24 hour simulated transportation in 140 $mg.l^{-1}$ carazolol than in those treated with 15 $mg.l^{-1}$ MS222.

Murai *et al.* (1979) used orally administered Valium (diazepam) at 0.04 $mg.kg^{-1}$ to enhance survival during transportation of American shad, *Alosa sapidissima*. As noted previously, oral administration has the problem that drugged food will not be consumed equally by all members of the group.

In summary

The principal issues in transportation are minimising the weight of water and equipment carried, minimising the costs of these materials and maximising the survival and health of the animals carried (Taylor and Solomon, 1979). With the exception of a few cases where animals can be transported out of water, maintenance of good water quality is the major factor in transportation and is the prime requirement for success. Some countries have standards for transportation and in Germany the levels of certain water quality factors, the maximum fish density and duration of the journey are controlled by legislation. While not always necessary, some degree of sedation during transportation can be a useful tool to assist in achieving these objectives.

References

Alvarez-Lajonchere, L. and Garcia-Moreno, B. (1982). Effects of some anaesthetics on postlarvae of *Mugil trichodon* Poey (Pisces, Mugilidae) for their transportation. *Aquaculture*. 28 (3-4): 385-390.

Anon (1982). Eel exports. *Fish Farmer*. 5 (4): 29.

Barton, B.A. and Helfrich, H. (1981). Time-dose response of juvenile Rainbow trout to 2-phenoxyethanol. *Progressive Fish Culturist*. 46 (4): 223.

Bell, G.R. (1964). A guide to the properties, characteristics and uses of some general anaesthetics for fish. *Bulletin of the Fisheries Research Board of Canada*. No. 148.

Berka, R. (1986). The transport of live fish. A review. *EIFAC Technical paper.* No. 48. CECPI, FAO, Rome, Italy. 52pp.

Chittenden, M.E. (1992). Transporting and handling young American shad. *New York Fish and Game Journal.* 18: 123-128.

Durve, V.S. (1975). Anaesthetics in the transport of mullet seed. *Aquaculture.* 5: 53-63.

Ferreira, J.T., Schoonbee, H.J. and Smit, G.L. (1984). The use of benzocaine-hydrochloride as an aid in the transport of fish. *Aquaculture.* 42 (2): 69-174.
Gopinath Nagaraj (1990). Biotechnical considerations in the handling and transport of live fishery products. In: *Handling and Processing: The Selling Point in Malaysia Fisheries* (Edited by N. Gopinath and A. Abu-Baker). Occasional publication No. 3. Malaysian Fisheries Society. Kuala Lumpur, Malaysia.

Guo, F.C., Teo, L.H. and Chen, T.W. (1993). Application of anaesthetics in the transport of platyfish by air. In: *From Discovery to Commercialisation.* (Edited by M. Carrillo, L. Dahle, J. Morales, P. Sorgeloos, N. Svennevig and J. Wyban). Special publication of the European Aquaculture Society. No. 19. 228pp.

Hovgaard, P. (1984). Culture of blue mussel in Ireland. *Norsk Fiskeoppdrettt.* 9 (6): 38.

Horvath, L., Tamas, G. and Tolg, I. (1984). *Special Methods in Pond Fish Husbandry.* Halver Corporation, Seattle. 146pp.

Innes-Taylor, N. and Ross, L.G. (1988). The use of hydrogen peroxide as a source of oxygen for the transportation of live fish. *Aquaculture.* 70 (1-2): 183-192.

Itazawa, Y. and Takeda, T. (1982). Respiration of carp under anesthesia induced by mixed bubbling of carbon dioxide and oxygen. *Bulletin of the Japanese Society of Scientific Fisheries.* 48 (4): 489-493.

Jhingran, V.G. and Pullin, R.S.V. (1985). *A Hatchery Manual for the Common, Chinese and Indian Major Carps.* Asian Development Bank/ICLARM, Manila, Philippines. 191pp.

Jirasek, J., Adamek, Z. and Giang, P.M. (1978). The effect of administration of the anaesthetics MS222 (Sandoz) and R7464 (Propoxate) on oxygen consumption in the tench (*Tinca tinca.* L.). *Zivocisna-Vyroba.* 23 (11): 835-840.

Johnson, S.K. (1979). Transport of live fish. *Aquaculture Magazine.* 5 (6): 20-24.

Kacem, N.H., Aldrin, J.F. and Romestand, B. (1988). Modifications de certaines reponses secondaires de stress chez *Dicentrarchus labrax* sous l'effect d'un beta-bloquant a base de Carazolol. Application au transport. *Aquaculture.* 68 (3): 277-285.

Laird, L.M. and Wilson, A.R. (1979). A method for improving the survival of fish eggs during transportation. *Fisheries Management*. 10 (3): 129-131.

Malyukina, G.A., Martem'-yanov, V.I. and Flerova, G.I. (1982). The alarm pheromone as a stress factor for fish. *Journal of Ichthyology*. 22 (2): 147-150.

McFarland, W.N. (1960). The use of anaesthetics for the handling and the transport of fishes. *California Fish and Game*. 46: 407-431.

Marathe, V.B., Huilgol, N.V. and Patil, S.G. (1975). Hydrogen peroxide as a source of oxygen supply in the transport of fish fry. *Progressive Fish Culturist*. 37 (2): 117.

Meeren, G. van der. (1991). *Acclimatisation Techniques for Lobster*. Norwegian Society for Aquaculture Research. Bergen, Norway. pp.95-97.

Meeren, G. van der and Uglem, I. (1993). Techniques for transportation and release of juvenile lobsters (*Homarus gammarus*). *Fisken Havet*. 7: 31.

Murai, T., Andrews, J.W. and Muller, J.W. (1979). Fingerling American shad: effect of valium, MS-222, and sodium chloride on handling mortality. *Progressive Fish Culturist*. 41 (1): 27-29.

Nakamura, K. (1992). Effect of precooling on cold air tolerance of the carp *Cyprinus carpio* . *Bulletin of the Japanese Society of Scientific Fisheries*. 58 (9): 1615-1620.

Okoye, R.N. (1982). Techniques for transportation of juvenile tilapia. M.Sc. Thesis, University of Stirling. 39pp.

Panggabean, L., Waterstrat, P.R., Downing, L. and Beattie, J.L. (1989). Storage and transportation of straight-hinge oyster larvae. *Journal of Shellfish Research*. 8 (1): 323-324.

Paterson, B.D. (1993). Respiration rate of the kuruma prawn, *Penaeus japonicus* Bate, is not increased by handling at low temperature (12°C). *Aquaculture*. 114 (3-4): 229-235.

Ross, R.M., Backman, T.W.H. and Bennett, R.M. (1993). Evaluation of the anesthetic metomidate for the handling and transport of juvenile American shad. *Progressive Fish Culturist*. 55 (4): 236-243.

Rothbard, S. (1988). The use of chemical tranquillisers in fish transport. *Fish and Fishbreeding in Israel*. 21 (1): 28-34.

Sado, E.K. (1985). Influence of the anaesthetic quinaldine on some tilapias. *Aquaculture*. 46 (1): 55-62.

Sampson, D.R.T. and Macintosh, D.J. (1986). Transportation of live carp fry in sealed polythene bags. *Aquaculture*. 54 (1-2): 123-127.

Singh, H., Chowdhury, A.R. and Pakrasi, B.B. (1982). Experiments on the transport of postlarvae of tiger prawn *Penaeus monodon* Fabricius. *Symposium Series of the Marine Biological Association of India*. No. 6. 232-235

Smith, T.I.J. and Ribelin, B. (1984). Stocking density effects on survival of shipped postlarval shrimp. *Progressive Fish Culturist*. 46 (1): 47-50.

Smith, T.I.J. and Wannamaker, A.J. (1983). Shipping studies with juvenile and adult Malaysian prawns *Macrobrachium rosenbergii* (de Man). *Aquaculture Engineering*. 2 (4): 287-300.

Solis, E.P. and Heslinga, G.A. (1989). Effect of desiccation on *Tridacna derasa* seed: Pure oxygen improves survival during transport. *Aquaculture*. 76 (1-2): 169-172.
Solomon, D.J. and Hawkins, A.D. (1981). Fish capture and transport. In: *Aquarium Systems* (Edited by A.D. Hawkins). Academic Press, London.

Takeda, T. and Itazawa, Y. (1983). Possibility of applying anesthesia by carbon dioxide in the transportation of live fish. *Bulletin of the Japanese Society of Scientific Fisheries*. 49 (5): 725-731.

Taylor, A.L. and Solomon, D.J. (1979). Critical factors in the transport of living freshwater fishes. I. General considerations and atmospheric gases. *Fisheries Management*. 10: 27-32.

Teo, L.H. and Chen, T.W. (1993). A study of metabolic rates of *Poecilia reticulata* Peters under different conditions. *Aquaculture and Fisheries Management*. 24 (1): 109-117.

Wedermeyer, G. (1972). Some physiological consequences of handling stress in the juvenile Coho salmon (*Onchorhynchus kisutch*). *Journal of the Fisheries Research Board of Canada*. 29: 1780-1783.

Weirich, C.R. Tomasso, J.R. and Smith, T.I.J. (1992). Confinement and transport-induced stress in white bass *Morone chrysops* x striped bass *M. saxatilis* hybrids: Effect of calcium and salinity. *Journal of the World Aquaculture Society*. 23 (1): 49-57.

Whiteley, E.M. and Taylor, E.W. (1992). Oxygen and acid-base disturbances in the hemolymph of the lobster *Homarus gammarus* during commercial transport and storage. *Journal of Crustacean Biology*. 12 (1): 19-30.

Woynarovich, E. and Horvath, L. (1980). The artificial propagation of warmwater fishes. A manual for extension. *FAO Fisheries Technical Paper*. No. 201. pp.138-147.

Yokoyama, Y., Yoshikawa, H., Ueno, S. and Mitsuda, H. (1989). Application of CO_2 anesthesia combined with low temperature for long-term anesthesia in carp. *Bulletin of the Japanese Society of Scientific Fisheries*. 55 (7): 1203-1209.

Yoshikawa, H., Ueno, S. and Mitsuda, H. (1989). Short- and long-term cold-anesthesia in carp. *Bulletin of the Japanese Society of Scientific Fisheries*. 55 (3): 491-498.

~~~~

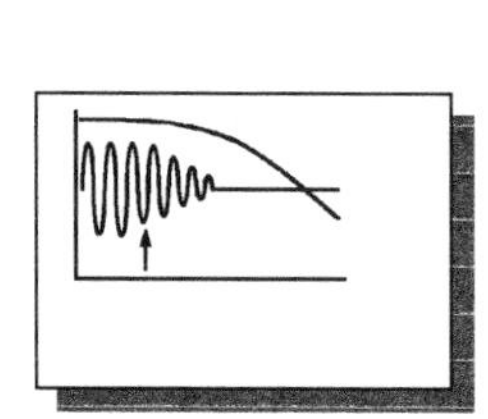
~~~~

Chapter 13
Concluding Remarks

In any given situation it is always worth reconsidering whether anaesthesia or sedation are strictly necessary. In Chapter 1 a number of instances have been given where anaesthesia may not be essential or where there are indeed contraindications to its use. Simple fish holders can be constructed which will constrain small specimens quietly, usually for some tens of seconds, allowing rapid tagging or marking. Many broodstock fish can be stripped without recourse to anaesthetics and without damage, so long as the handler is gentle but firm. As a general rule it is advisable to avoid the use of chemicals with organisms destined for human consumption for obvious reasons. Thus, although the US Food and Drugs Administration (FDA) allows the use of MS222 and quinaldine with cultured fish, they require that no drugs should be used for 21 days prior to sacrifice.

Having decided that some degree of immobilisation is desirable, the selection of a suitable anaesthetic technique will be influenced by the situation in which it will be used and thus no simple guidelines can be given. As this book attempts to show, there are advantages and disadvantages inherent in all of the methods described and these should be weighed against the requirements of the work in hand. In addition to availability of equipment, the availability and cost of drugs may limit the choices. It would be prudent in most cases to have more than one drug available, for example using benzocaine as a routine but holding small stocks of MS222 for use when no appropriate organic solvent is available. Field workers may note that judicious use of vodka or gin with benzocaine can overcome solvent problems in emergencies.

Many workers rely on one drug, presumably adhering to the old axiom "better the devil you know than the one you don't". This will serve well in most instances but, again, the pros and cons of all the available methods should be weighed against the objectives of the work. A single approach will not serve for all ends and it is hoped that this book, at the least, may engender some flexibility.

There are considerable practical advantages to be gained by using electrical anaesthetic techniques. Following modest initial outlay, running costs are relatively low. It should be borne in mind that anaesthetic drugs are frequently expensive, or difficult to obtain in certain parts of the world, whereas electricity will be available or can be made available by the use of inexpensive portable generators. A problem with electrical anaesthesia is that of operator safety and while the use of isolating transformers, circuit breakers and interlocking switches can reduce the dangers, it is virtually impossible to guarantee that electric shocks cannot be sustained. It is therefore most important to adhere to a well-considered code of practice and although the technique for practical implementation has been described here, it should be noted that the authors accept no responsibility for accidents in the use of this equipment, howsoever caused.

It is possible that further research will bring electroanaesthesia into wider use in aquaculture, but at present drugs, notably benzocaine, MS222 and quinaldine, are used routinely world-wide for most purposes. For more complex procedures and surgical work chemicals are, and will probably remain, the first choice as much more is known of the physiological effects. The tables provided throughout the book are an attempt to draw together some of the available data on the major drugs in current use, in order to provide initial guidelines for inexperienced and experienced operators alike. As further investigations proceed, and researchers and growers test new materials and approaches, the tables could be enhanced, thereby becoming infinitely more useful. As we noted in the first edition, we would be interested to receive communications regarding problems and experiences from our colleagues, old and new, world-wide.

~~~~

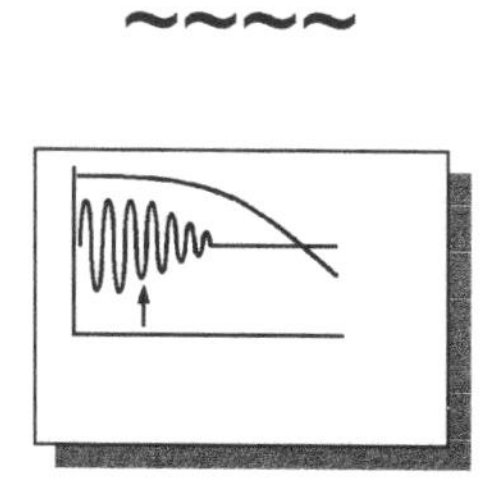
~~~~

Glossary of Nomenclature and Structure of Common Anaesthetic Drugs

Generic or common name	Alternative names, trade names	Chemical name, molecular formula **Structural formulae in square brackets [ ]**
2-amino-4-phenylthiazole	APT, phenthiazamine, phenthiazamine hydrobromide, Piscaine	2-amino-4-phenylthiazole hydrobromide monohydrate
2-phenoxyethanol	phenoxethol, chlorophenoxetol	1-hydroxy-2-phenoxyethane $C_8H_{10}O_2$ [$C_6H_5OCH_2CH_2OH$]
4-styrylpyridine	–	4-styrylpyridine
alphadolone	a steroid anaesthetic	3α,21-dihydroxy-5α-pregnane-11,20-dione-21 acetate
alphaxolone	a steroid anaesthetic	3α-hydroxy-5α-pregnane-11,20-dione
acetylpromazine	acepromazine, ACP, Plegicil	3-acetylpromazine
barbiturates	Amytal, amobarbital, amylobarbitone, sodium amylobarbitone	5-ethyl-5-isoamylbarbituric acid sodium salt $C_{11}H_{17}N_2O_3Na$
	Brevital, Brietal sodium methohexitone	1-methyl,5-allyl1,5-(1-methylpentynyl) barbituric acid
	hexobarbitone, Evipan, Evipal	5-(1-cyclohexenyl)-1,5-dimethyl-barbituric acid
	Nembutal, sodium pentobarbital, pentobarbitone, Euthatal, Exital, numerous other trade names	5-ethyl-5-(1-methylbutyl) barbituric acid sodium salt $C_{11}H_{17}N_2O_3Na$
	phenobarbitone, phenobarbital	5-ethyl-5-phenyl barbituric acid
	seccobarbital, sodium thiamylal, quinalbarbitone, Seconal	5-allyl-5-[1-methylbutyl]barbituric acid sodium salt $C_{12}H_{17}N_2O_3Na$
	thiopentone, Pentothal, Intraval, thiopental, thiopentone sodium, Thiovet	5-ethyl-5-(1-methylbutyl)-2-barbituric acid
benzocaine	ethyl-p-aminobenzoate, 4-aminobenzoic acid ethyl ester	ethyl-4-aminobenzoate $C_9H_{11}NO_2$ [$H_2NC_6H_4CO_2C_2H_5$]
butyl alcohol	n-butyl alcohol	1-butanol C_4H_9OH [CH_3-CH_2-CH_2-CH_2-OH]
carbon dioxide	dry ice (as a solid)	carbon dioxide (CO_2)

chloral hydrate	–	chloral hydrate $CCl_3CH(OH)_3$
chlorbutanol	chloretone, chlorbutol β,β,β-trichloro-t-butanol	1,1,1-trichloro-2-methyl-2-propanol $C_4H_7ClO_3.2H_2O$
chlordiazepoxide	Librium	7-chloro-2-(methylamino)-5-phenyl-3H-1,4-bezodiazepine 4-oxide
chloroform	–	chloroform $CHCl_3$
decamethonium (chloride)	–	decamethylenebis (trimethyl-ammonium chloride) $(CH_3)_3.N(CH_2)_{10}N.(CH_3)_3$
diazepam	Valium	a benzodiazepine
diethyl ether	ether, ethyl ether	diethyl ether $(CH_3CH_2)_2O$
diprenorphine	see Revivon	–
enflurane	Enthrane	2-chloro-1,1,2-trifluoroethyl difluoromethyl ether $CHF_2.O.CF_2.CHFCl$
ethyl alcohol	alcohol, ethanol	ethanol C_2H_5OH $[CH_3\text{-}CH_2\text{-}OH]$
etomidate	Hypnomidate, Amidate	an imidazole derivative
etorphine	M99, see Immobilon	an oripavine derivative of the morphine molecule
eugenol	principal oil extracted from clove oil	4-allyl-2-methoxyphenol
fentanyl	Sublimaze, Hypnorm	N-phenyl-N-(1-(2-phenylethyl)-4-piperidinyl) propanamide
Fluothane	see halothane	see halothane
gallamine	Flaxedil	gallamine triethiodide $C_6H_3\text{-}1,2,3\text{-}(OCH_2CH_2N(C_2H_5)_3I)_3$
halothane	Fluothane	2-bromo-2-chloro-1,1,1-trifluoroethane $CHBrCl.CF_3$ C_2F_3HBrCl
heptanol	heptyl alcohol	1-heptanol $C_7H_{16}O$ $[CH_3(CH_2)_6OH]$
Immobilon	etorphine+acetylpromazine etorphine+methotrimeprazine	= large animal Immobilon = small animal Immobilon
isoflurane	Forane	1-chloro-2,2,2-triflouroethyl difluoromethyl ether $CHF_2.O.CClF.CHF_2$
ketamine	ketamine hydrochloride, Ketaset, Vetalar, Ketalar, related to phencyclidine	2-(2-chlorophenyl)-2-(methylamino) cyclohexanone $ClC_6H_4C_6H_8(=O)NHCH_3.HCl$

lignocaine	Lidocaine, Lignavet, Nopoane, Xylocaine, Xylotox	2-diethylamino-N-[2,6-dimethylphenyl acetamide] hydrochloride $C_{14}H_{22}N_2O$
menocain	–	sodium hydrogen sulphate ethyl 3-amino benzoate
methoxyfluorane	Penthrane, Metofane	2,2-dichloro-1,1-difluoroethyl methyl ether $CHCl_2CFOCH_3$
methyl pentynol	methyl parafynol, Dormisan	$CH_3.CH_2.CCH_3OH.COH$
metomidate	Hypnodil, Methoxymol, Marinil, R7315	di-1-(1-phenylethyl)-5-(methoxycarbonyl) imidazole hydrochloride
MS222	ethyl-m-aminobenzoate, ethyl-3-aminobenzoate, methanesulphonic acid, tricaine, tricane methane sulphonate, Finquel	2-methyl 4-sulpholnyl aminobenzoate $C_9H_{11}NO_2.CH_4SO_4$ $[H_2NC_6H_4CO_2C_2H_5.CH_3SO_3H]$
nalorphine	lethiotone	antagonist for morphine and derivatives
Novocaine	see procaine	–
phencyclidine	–	1-(1-phencyclohexyl) piperidine
Piscaine	see 2-amino-4-phenylthiazole	see 2-amino-4-phenylthiazole
procaine (hydrochloride)	Novocaine, Planocaine, Corneocaine	2-(diethylamino)ethyl 4-aminobenzoate (hydrochloride) $[H_2NC_6H_4CO_2CH_2.CH_2N(C_2H_5)_2.HCl]$
propanidid	Epontol, fabontol, fabontal. a eugenol derivative	3-methyl-4(NN diethyl carbamoyl methoxy) phenylacetic acid n-propyl ester propyl-4-diethylcarbamoylmethoxy-3-methoxyphenylacetate
propoxate	propoxate hydrochloride, R7464	propyl-di-(1-phenylethyl)imidazole-5-carboxylate hydrochloride
quinaldine	quinaldin	2-methyl quinolone $C_{10}H_9N$
quinaldine sulphate	–	2-methyl quinolone sulphate $C_{10}H_9N.H_2SO_4$
Revivon	diprenorphine-HCl	based on the morphine molecule
Rompun	xylazine-HCl	see xylazine
Saffan	althesin, alphaxolone+alphadolone 3:1	see alphaxolone and alphadolone

suxamethonium chloride	Anectin, Sucostrin, scoline	$3(CH_3).N(CH_2).COO.(CH_2)_2.COO.$ $(CH_2)_2N.3(CH_3)$
telazol	CI-744, tiletamine hydrochloride + zolazepam hydrochloride	see tiletamine
tertiary amyl alcohol	TAA, amylene hydrate	2-methyl 2-butanol $C_5H_{12}O$
tertiary butyl alcohol	TBA, t-butanol	$C_4H_{10}O$
thiouracil	4-hydroxy-2-mercaptopyrimidine	2-thiouracil
tiletamine (hydrochloride)	CI-634	a benzodiazepine
tribromoethanol	Avertin, tribromethanol	tribromoethanol $CBr3CH_2OH$
tricaine	see MS222	see MS222
urethane	ethyl carbamate, urethan	ethyl carbamate $C_3H_7NO_2$ $[H_2NCO_2C_2H_5]$
Vetalar	see ketamine-HCl	see ketamine
xylazine	Rompun, Anased, Vibraxyl	N-(5,6-dihydro-4H-1,3-thiazin-2-yl)-2,6-xylidine-
Xylocaine	see lignocaine	–

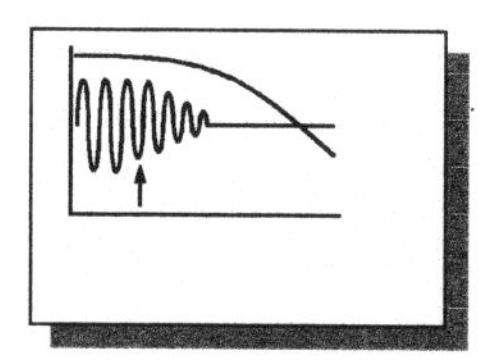

Glossary of Technical Terms

Acute	Having a rapid effect, and not long-lasting (opposite to chronic)
Alternating current, A.C.	An electrical current whose voltage regularly varies (alternates) from a positive to negative level. Alternating currents (A.C.) may have a sinusoidal, triangular or square waveform. Domestic mains alternating currents are sinusoidal
Analgesia	Relief from pain
Anaesthesia Anesthesia (USA)	A reversible, generalised loss of sensory perception accompanied by a sleep-like state induced by drugs or by physical means (see Chapter 4)
Anoxia	Exposure to zero environmental oxygen concentration
Ataxia	Uncoordinated movements
Biopsy	Removal of a sample of live tissue (e.g. liver, spleen) for biochemical or pathological analysis
Bradycardia	Reduced heart rate
Branding	A form of medium-term marking of animals using heat or cold damage to the skin
Catecholamines	Hormones released by the interrenal bodies (in fish) and having widespread metabolic activity. Also responsible for transmission of nervous signals at synapses
Chopped D.C.	An alternating waveform with a rapid rise and fall time, usually varying from 0 V to some positive voltage, created by switching of a D.C. source. Widely used in electric fishing
Chromatophore	A cell in the integument (skin) containing dispersible or concentratable pigment in one of a range of colours
Chronic	Long-lasting (relatively), opposite to acute
Cough response	Momentary reversal of water flow over the gills. Almost all fish species exhibit this function
Direct current, D.C.	An electrical current whose voltage is at a fixed positive or negative level and which does not vary regularly
Drug	A chemical substance which when consumed by, or applied to, an animal has certain biological effects
Ectotherm	An animal whose body temperature is not controlled and closely follows that of the environment, e.g. invertebrates, fish, amphibia and reptiles

ECG (Electrocardiogram)	An electrical record of heart muscle contractions usually recorded from surface electrodes applied to an animal
Electroanaesthesia	Anaesthesia brought about by electrical stimulation of an animal, or part of an animal. Some workers contest that this is "true" anaesthesia
Electronarcosis	As electroanaesthesia
Galvonarcosis	Electroanesthesia brought about by use of a direct current (D.C.). Its effects are usually apparent only when current is applied
Gonadectomy	Surgical removal of the gonads
Haemoconcentration	Concentration of the blood by loss of water, gain of ions or addition of cells
Haemodilution	The converse of haemoconcentration
Hypophysectomy	Surgical removal of the pituitary gland
Hypoxia	Exposure to low environmental oxygen concentration
Intramuscular (IM)	Injection given into the muscle
Intraperitoneal (IP)	Injection given into the peritoneum, the body cavity
Local anaesthesia	Affecting a localised area, e.g. an area of skin
$mg.l^{-1}$	Milligrams of a solute per litre of solvent
Narcosis	As for anaesthesia, but does not necessarily presume that recovery will occur
Narcotic	A sleep inducing agent
Osmolarity	The solute load (e.g. of blood) in kg/litre
ppm	Parts per million. Usually refers to a solution and measured in µl of a liquid per litre of solvent (v/v) or mg of a solid per litre of solvent (w/v)
Pulsed D.C.	See chopped D.C.
Sedation	A preliminary level of anaesthesia, in which response to stiimulation is greatly reduced and some analgesia is achieved, but sensory abilities are generally intact and loss of equilibrium does not occur
Square wave	An alternating waveform with a rapid rise and fall time. Can be generated by chopping of a D.C source or by adding full harmonics to a sine wave
Steroids	A group of powerful hormones having widespread metabolic actions. Released by the interrenal bodies under control of the hypothalamus/pituitary
Stressor	Any environmental variable exceeding normal bounds
Tachycardia	Accelerated heart rate
Tachyventilation	Accelerated gill ventilation rate (movement of opercula in fish)

Taxis	Motor activity induced by external stimulation
Topical anaesthesia	Affecting a localised area, e.g. an area of skin
Telemetry	Remote sensing of data by radio or ultrasonic means, in this case from an animal

~~~~

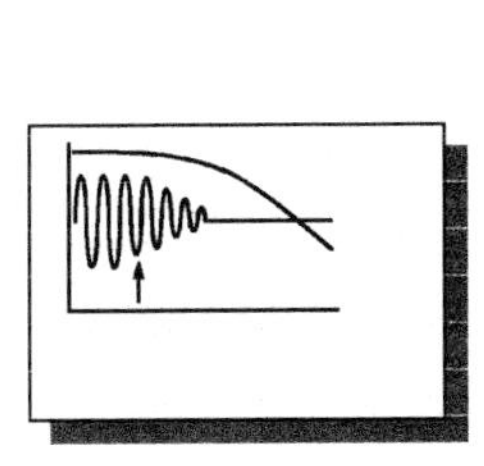
~~~~

Index